The Separable Galois Theory of Commutative Rings

PURE AND APPLIED MATHEMATICS

A Series of Monographs and Textbooks

1. K. YANO. Integral Formulas in Riemannian Geometry (1970)
2. S. KOBAYASHI. Hyperbolic Manifolds and Holomorphic Mappings (1970)
3. V. S. VLADIMIROV. Equations of Mathematical Physics (A. Jeffrey, editor; A. Littlewood, translator) (1970)
4. B. N. PSHENICHNYI. Necessary Conditions for an Extremum (L. Neustadt, translation editor; K. Makowski, translator) (1971)
5. L. NARICI, E. BECKENSTEIN, and G. BACHMAN, Functional Analysis and Valuation Theory (1971)
6. D. S. PASSMAN. Infinite Group Rings (1971)
7. L. DORNHOFF. Group Representation Theory (in two parts). Part A: Ordinary Representation Theory. Part B: Modular Representation Theory (1971, 1972)
8. W. BOOTHBY and G. L. WEISS (eds.). Symmetric Spaces: Short Courses Presented at Washington University (1972)
9. Y. MATSUSHIMA. Differentiable Manifolds (E. T. Kobayashi, translator) (1972)
10. L. E. WARD, JR. Topology: An Outline for a First Course (1972)
11. A. BABAKHANIAN. Cohomological Methods in Group Theory (1972)
12. R. GILMER. Multiplicative Ideal Theory (1972)
13. J. YEH. Stochastic Processes and the Wiener Integral (1973)
14. J. BARROS-NETO. Introduction to the Theory of Distributions (1973)
15. R. LARSEN. Functional Analysis: An Introduction (1973)
16. K. YANO and S. ISHIHARA. Tangent and Cotangent Bundles: Differential Geometry (1973)
17. C. PROCESI. Rings with Polynomial Identities (1973)
18. R. HERMANN. Geometry, Physics, and Systems (1973)
19. N. R. WALLACH. Harmonic Analysis on Homogeneous Spaces (1973)
20. J. DIEUDONNÉ. Introduction to the Theory of Formal Groups (1973)
21. I. VAISMAN. Cohomology and Differential Forms (1973)
22. B.-Y. CHEN. Geometry of Submanifolds (1973)
23. M. MARCUS. Finite Dimensional Multilinear Algebra (in two parts) (1973)
24. R. LARSEN. Banach Algebras: An Introduction (1973)
25. R. O. KUJALA and A. L. VITTER (eds.). Value Distribution Theory: Part A; Part B, Deficit and Bezout Estimates by Wilhelm Stoll (1973)
26. K. B. STOLARSKY. Algebraic Numbers and Diophantine Approximation *(in preparation)*
27. A. R. MAGID. The Separable Galois Theory of Commutative Rings (1974)

The Separable Galois Theory of Commutative Rings

Andy R. Magid
Department of Mathematics
University of Oklahoma
Norman, Oklahoma

MARCEL DEKKER, INC. New York 1974

MARCEL DEKKER, INC.

305 East 45th Street, New York, New York 10017

LIBRARY OF CONGRESS CATALOG CARD NUMBER: 74-80286

ISBN: 0-8247-6143-4

Current printing (last digit):

10 9 8 7 6 5 4 3 2 1

PRINTED IN THE UNITED STATES OF AMERICA

To Carol

ACKNOWLEDGMENTS

Among the many colleagues and friends whose assistance was most helpful in the development of the material in this volume, I am particularly grateful to Daniel Zelinsky, Lindsay Childs, and Raymond Hoobler for their enlightenment and guidance.

It is also a pleasure to be able to thank Trish Abolins for her able and efficient typing of this volume.

A.M. Norman, 1973

CONTENTS

Introduction ix

I. Profinite Topological Spaces........................ 1

II. The Boolean Spectrum.............................. 23

III. Galois Theory over a Connected Base............... 45

IV. The Fundamental Groupoid.......................... 77

V. Galois Correspondences............................ 111

Bibliography.. 131

Index 133

Introduction

These notes are intended to present the recent developments in the theory of commutative separable algebras over commutative rings in a reasonably accessible form. We actually have three purposes in mind: first, the theory itself, while not particularly complicated, draws heavily on two topics (totally disconnected compact Hausdorff topological spaces and the Pierce-Villamayor-Zelinsky Boolean spectrum of a commutative ring) which are not usually a part of standard algebra. We've tried, therefore, to provide a relatively complete and self-contained introduction to those aspects of these topics which are involved in the theory. Second, as the theory has been discovered and expanded certain concepts have been enlarged and extended, so that the same idea with a different definition (or the same definition but with reference to a different idea) appears in various places in the literature - this applies in particular to the notion of separable closure. So another purpose of these notes is to clarify and (hopefully) standardize the language of the theory. Third, the theory itself has appeared in print previously in the chronological order of its discovery, rather than in logical order, and this has obscured the independence of the results. Now that the theory is more or less complete, we have tried to reorganize it in such a way that the major themes of the subject appear in their proper order.

There are two such major themes to the Galois theory of separable algebras over rings: one leading to the correspondence theorem, which attempts to describe separable subalgebras of a separable algebra in terms of the subobjects of some algebraic object attached to the algebra (if the base ring is a field and the algebra a normal, separable finite dimensional extension field then the fundamental theorem of the Galois theory of fields says that there is a one-one correspondence between sub-extension fields and subgroups of the group of all automorphisms of the extension field which leave the elements of the base field fixed), and the other to the classification theorem, which attempts to describe the category of all separable algebras in terms of some other category (if the base ring is a field, then the category of all finite separable extension fields is anti-equivalent, via the Krull-Galois theory of infinite extensions, to the category of finite sets on which the (topological) group of all automorphisms of the separable closure of the base field fixing the base field acts continuously and transitively).

The two themes are obviously intertwined. In the case of fields, for example, the correspondence theorem plays a part in proving the classification theorem, and in turn, given the classification theorem, it's an easy matter to prove a correspondence theorem in the category of sets with group action, and then use classification to deduce a correspondence theorem for extension fields.

There is, however, some advantage in viewing the classification theorem as basic and the correspondence theorem as an application, and this is the attitude adopted here. We hope this enables the reader to see the correspondence theorem, despite its name (Fundamental Theorem of Galois Theory), as only one application of the classification theorem.

2.

Much research has been devoted to discovering the proper setting in which to extend the Galois theory of fields to commutative rings, and we'll outline here only a very brief account of its development. The appropriate definitions were introduced by Auslander and Goldman in "The Brauer group of a commutative ring", Trans. Amer. Math. Soc. 97 (1960), and the correspondence theorem explored by Chase, Harrison, and Rosenberg in their Amer. Math. Soc. Memoir #52, "Galois theory and Galois cohomology of commutative rings". Because of their concern with cohomology, these theories used groups of automorphisms, and their correspondence theorem, while it included all the subgroups, included only certain subalgebras, unless the rings involved had no idempotents except zero and one. Villamayor and Zelinsky, in "Galois theory for rings with finitely many idempotents", Nagoya Math. J. 27(1966), were able to give a complete correspondence in the case of their title, but using groupoids instead of groups. Then in "Galois theory for rings with infinitely many idempotents", Nagoya Math. J. 35(1969), they were able to drop their hypotheses on idempotents, but topological considerations forced them to return to groups, and this made the correspondence again incomplete: it included all the separable subalgebras but only certain subgroups. Their results were extended to infinite algebras in "Locally Galois algebras", Pacific J. Math. 33(1970). Then a way was discovered to reintroduce groupoids, and a complete correspondence, which worked for infinite as well as finite algebras, was given in "Galois groupoids", J. of Algebra 18(1971). Thus the correspondence theorem developed independently of the classification theorem.

The classification theorem was given by Grothendieck in the _Seminaire Geometrie Algebrique_, 1960-61, Institute des Hautes Etudes Scientifiques, in the language of schemes, for separable algebras over rings with no idempotents except zero and one. In order to extend this to rings with arbitrary idempotents, a suitable notion of separable closure was required. This was first investigated in "The separable closure of some commutative rings", Trans. Amer. Math. Soc. 170(1972), where the proper algebras to be classified appeared, and the classification theorem itself was obtained in "The fundamental groupoid of a commutative ring".

3.

The notes are divided into five chapters, which we now rapidly summarize: Chapter I, _Profinite Topological Spaces_, lays down the topological foundations necessary for the later work. After giving the various equivalent formulations of the notion of profinite space, extremely disconnected spaces are introduced, and characterized. The (Gleason) minimal extremely disconnected cover of a space is constructed. Then profinite group actions on profinite spaces are examined, and the chapter concludes with some facts about rings of locally constant functions on profinite spaces.

Chapter II, The Boolean Spectrum, is about the space of connected components of the spectrum of a commutative ring and the canonical sheaf that space supports. First it is shown that the topology of the space of components determines and is determined by the idempotents of the ring. Then a number of "lifting" theorems are proved: results about whether algebraic data defined over the stalks of the sheaf on the space of components can be lifted to global data defined over the whole sheaf and hence over the ring; this can be done for any data of finite type, essentially. Some connections with the rings of locally constant functions of Chapter I are studied. (Throughout Chapter II, we use an approach which avoids the sheaf language for ease of exposition so the description just given is not exactly accurate.)

The Boolean spectrum is studied because it provides a method of reducing some problems about general rings to similar problems about rings with no non-trivial idempotents. Chapter III, Galois Theory Over A Connected Base, studies locally strongly separable* algebras

*Following standard usage 'strongly' means 'module-projective' and 'locally' means 'direct limit of'.

over a ring with no idempotents except zero and one. The first part of the chapter is a construction of the separable closure from the top down (the usual construction is from the bottom up, but we want to set the correct pattern for the later construction over arbitrary base rings). The concept of normality is introduced, under the name 'weakly Galois', and the basic structure of locally weakly Galois algebras is determined - they're all rings of locally constant functions, on profinite spaces, with values in (possibly infinite) Galois algebras. This structure theorem is used to prove a Galois correspondence theorem: there's a one-one correspondence between locally strongly separable subalgebras of a locally weakly Galois algebra and the closed subgroupoids of a certain topological groupoid.

Chapter IV, The Fundamental Groupoid, looks at the problem of carrying out the construction of the separable closure in the general case, i.e., where no conditions are placed on the idempotents in the base ring. This requires a study of componentially* locally strongly separable algebras (previously called quasi-separable covers). The separable closure is such an algebra which receives homomorphic images of all such algebras and is minimal with respect to this property. It's constructed by first adding idempotents to get an algebra whose Boolean spectrum is the Gleason cover of the Boolean spectrum of the base ring and then separably closing, in the sense of rings with no non-trivial

---

*'Componentially' means 'stalkwise' and is applied to sheaves on the space of components of the spectrum of a ring.

idempotents, each of the stalks of the canonical sheaf on the Boolean spectrum of this algebra. The category of componentially locally strongly separable algebras over the separable closure is shown to be anti-equivalent to a category of profinite spaces. The corresponding category of algebras over the base ring, then, is described by the classification theorem: it's anti-equivalent to a category of profinite spaces on which a certain topological groupoid acts. This last result uses nearly everything in the earlier chapters, and is the main point of the theory.

Chapter V, Galois Correspondences, carries out the program mentioned above: the Galois correspondence theorem for componentially locally strongly separable algebras over an arbitrary base ring is proved from the analogous theorem in a category of profinite spaces with groupoid-action. First the general correspondence between equivalence relations and quotients is studied in this latter category. This is translated into a correspondence between equivalence correlations and subobjects in the category of algebras. Under suitable normality conditions, this becomes: there is a one-one correspondence between componentially locally strongly separable subalgebras of a normal componentially locally strongly separable algebra and the closed subgroupoids of a certain topological groupoid. The chapter, and the notes, conclude with this form of the correspondence theorem.

4.

These notes, while they are intended to place in a convenient form the Galois theory of commutative rings discovered so far, are not, of course, the whole story. Among the current developments, there is the general theory of separable polynomials, due to DeMeyer, which uses the theory of the separable closure discussed in these notes to study separable algebras generated (as algebras) by a single element, and a theory, discussed in "Principal homogeneous spaces and Galois algebras" (to appear), which classifies separable algebras on which finite groups act in a nice way. These results, like the correspondence theorem, are applications of the classification theorem.

Among the things which still need to be done, we'll single out just two: the cohomology of sheaves in the Grothendieck topology generated by the componentially locally strongly separable algebras should be computable from groupoid cohomology in the same way that the corresponding cohomology over rings with no non-trivial idempotents is computable from group cohomology, and there should be corresponding theories of cohomological dimension. Also, the classification theorem should work for schemes as well as for rings. (There have been some successful efforts to extend the correspondence theorem to schemes, but of course these would be consequences of the successful extension of the classification theorem.)

5.

So that these notes will be comprehensible to beginning as well as practicing separable algebraists, efforts have been made to keep the notes self-contained, particularly in the early chapters. As a prerequisite, besides the standard ring theory-module theory and topology, a basic background in separable algebras is necessary. The best available source of this at present is DeMeyer and Ingraham's book *Separable Algebras over Commutative Rings* (Math Lecture Notes #181, Springer-Verlag, New York, 1971). When it's necessary or useful to refer to a result from basic separable algebra, the reader is referred to this book which we indicate throughout by DI. Other sources for the material we discuss are explained in the bibliographic notes appended to each chapter.

We observe standard conventions (rings have identities, modules are unitary, tensor products and direct sums and products are unindexed when no ambiguity arises, etc.), and one idiosyncrasy: *all our rings are commutative*, but occasionally for reasons of euphony we redundantly call a ring a commutative ring.

The Separable Galois Theory of Commutative Rings

I. Profinite Topological Spaces.

This chapter is a collection of some facts about profinite topological spaces (also known as zero-dimensional compact spaces or totally disconnected compact spaces) which will be used in later chapters. Most readers will want to skip this chapter for now, and refer back to it as the need arises.

Definition I.1. A topological space which is an inverse limit of finite discrete spaces will be called a profinite topological space.

A profinite space is compact and Hausdorff since it is an inverse limit of compact Hausdorff spaces. It also has some strong separation properties, which will first be defined and then established:

Definition I.2. Two subsets P and Q in a topological space X can be separated if there are disjoint open sets U and V of X, containing P and Q respectively, whose union is all of X.

Obviously only disjoint sets can be separated, and the sets which do the separating are both open and closed.

Lemma I.3. Let the topological space X be an inverse limit of discrete spaces. Then any two disjoint points p and q of X can be separated.

Proof: By hypothesis there is a discrete space D and a continuous map f from X to D such that $f(p) \neq f(q)$. Divide D into disjoint subsets E and F containing $f(p)$ and $f(q)$ respectively. Then $f^{-1}(E)$ and $f^{-1}(F)$ are open sets of X separating p and q.

Lemma I.4. Let X be a topological space, C and D compact subsets and suppose that any point of C and any point of D can be separated. Then C and D can be separated.

Proof: First suppose C is just a single point p. For each point q of F let $U(q)$ and $V(q)$ separate p from q. Then the $V(q)$'s cover F and since F is compact, a finite number will cover F. Let V be the union of these (finitely many) $V(q)$'s and let U be the intersection of the corresponding $U(q)$'s. Then U and V separate p and F. For a general C, separate each point of C from F and proceed as above using the compactness of C.

Proposition I.5. Let X be a compact space. The following are equivalent:

i) X is Hausdorff and has a basis of open-closed sets,

ii) any two distinct points in X can be separated.

Proof: Suppose X is Hausdorff and has a basis of open-closed sets. Let p and q be distinct points of X and let U be an open-closed subset containing one of the two points. Then U and its complement separate p and q. Conversely, suppose any two distinct points in X can be separated. Let U be a non-empty open set of X containing, say, the point p. By Lemma I.4, p can be separated from the complement of U, and the open-closed subset which contains p is contained in U, so X has a basis of open-closed sets. Clearly, X is Hausdorff.

Proposition I.5 applies in particular to profinite spaces. In fact it will be shown that spaces satisfying the equivalent conditions of I.5 are automatically profinite. Some additional language is necessary:

Definition I.6. A partition of a topological space is a finite family of disjoint open subsets of the space which covers it.

Lemma I.7. Let X be a compact space which has a basis of open-closed sets. Then every open cover of X has a refinement which is a partition.

Proof: Every open cover can be refined to a cover consisting of basic (i.e., open-closed) subsets, and by compactness the cover may be assumed to be finite. If the open-closed sets $F_1,\cdots,F_n$ cover X, then so do the disjoint open-closed sets $E_1,\cdots,E_n$ where $E_i = F_i - F_{i-1} - \cdots - F_1$.

Proposition I.8. The following are equivalent for a topological space X :

i) X is compact Hausdorff and has a basis of open-closed sets,

ii) X is profinite.

Proof: Suppose X is compact Hausdorff and has a basis of open-closed sets. The partitions of X are inversely ordered by refinement and for any partition P of X there is a continuous map from X to P which sends x in X to the subset of P to which x belongs. Thus if $\bar{X}$ denotes the inverse limit of the partitions there is a continuous map f from X to $\bar{X}$. By Proposition I.5 any two points in X can be separated so f is one-one and since X maps onto each partition the image is dense. X being compact and $\bar{X}$ Hausdorff implies then that f is onto, and since f is closed it is a homeomorphism. Suppose conversely that X is profinite. Then X is compact Hausdorff and Proposition I.5 and Lemma I.4 imply X has a basis of open-closed sets.

There's a third characterization of profinite spaces which is also convenient to have available. Recall that a topological space is *totally disconnected* if the connected components are all reduced to points. A topological space in which distinct points can be separated is obviously totally disconnected, so a profinite space is totally disconnected. It will turn out that a compact Hausdorff totally disconnected space is necessarily profinite.

Lemma I.9. Let X be a compact Hausdorff space, p a point of X and A the set of all points of X which can be separated from X. Then A is open with connected complement.

Proof (Hurewicz and Wallman): If q belongs to A there are disjoint open sets U and V containing p and q respectively whose union is X. Then any point of V is also in A, and hence V is an open neighborhood of q in A. Let B be the complement of A. If B is disconnected, there are disjoint subsets C and D of B, closed in B, whose union is B. Let p belong to, say, C. C and D are closed in X and since B is closed there is an open set U of X containing C but whose closure is disjoint from D. The boundary E of U is a closed set missing both C and D and hence contained in A. It follows by Lemma I.4 that E and p can be separated. Let V be an open-closed set containing E but not p. We have $p \in U\text{-}V$. Since V is closed $U\text{-}V$ is open and since the boundary of U belongs to V, $U\text{-}V$ is also the closure of U minus V, and hence $U\text{-}V$ is closed. So $U\text{-}V$ is an open-closed set containing p but missing D. This means every point of D is separated from p, so since D is contained in B, D is empty.

Proposition I.10. The following are equivalent for a topological space X:

i) X is profinite,

ii) X is compact Hausdorff and totally disconnected.

Proof: As has already been remarked, a profinite space is compact, Hausdorff, and totally disconnected. Lemma I.9 implies that any two distinct points in a totally disconnected compact Hausdorff space can be separated, so by Proposition I.5 the space has a basis of open-closed sets and so, by Proposition I.8 the space is profinite.

It's possible to have a compact totally disconnected space which is not Hausdorff, and such a space can't be profinite (for an example, take two copies of the convergent sequence $1, \frac{1}{2}, \frac{1}{3}, \cdots, 0$ and identify the non-zero points). So taking an arbitrary compact space and identifying connected components to points need not produce a profinite space. Since it will be necessary to construct spaces this way later on, conditions which produce a compact space will now be studied.

Definition I.11. Let X be a topological space. The set $\mathrm{Comp}(X)$ of connected components of X, equipped with the weakest topology making the obvious surjection $X \to \mathrm{Comp}(X)$ continuous is called the space of components of X.

Lemma I.12. For any topological space X the space $\mathrm{Comp}(X)$ is totally disconnected.

Proof: Let C be a component of $\mathrm{Comp}(X)$ and let the inverse image of C in X be the disjoint union of the closed sets E and F.

Since E and F consist o fdistinct components, their images in Comp(X) are disjoint and closed. Since these images union up to C either E or F must be empty. It follows that the inverse image of C is connected and C itself is one point.

<u>Proposition I.13</u>. Let X be a compact topological space. The space of components of X is Hausdorff if and only if every component of X is an intersection of open-closed sets.

<u>Proof</u>: If Comp(X) is Hausdorff, by Lemma I.12, Proposition I.11, and Proposition I.8 every point of Comp(X) is an intersection of open-closed sets. Taking inverse images, it follows that every component of X is an intersection of open-closed sets. Conversely, suppose every component of X is an intersection of open-closed sets. Let C and D be distinct components of X . C and D are compact, and any point of D can be separated from C . Thus C and D can be separated by open-closed sets U and V which are disjoint. U and V consist of components and hence have disjoint open images in Comp(X) , and these images separate the points C and D of Comp(X) .

The intersection of all open-closed subsets of a topological space X which contain the point p is called the quasi-component of p . Proposition I.13 says that if X is compact, Comp(X) is Hausdorff if and only if the quasi-components of X are connected. As an example of a space where the quasi-components are different from the components we can use the example referred to above just before Definition I.11. The two copies of 0 are in the same quasi-component but not in the same component.

Next, we look at the notion of a covering of a profinite space. When dealing with spaces having several components, it's sometimes convenient to replace the concept of covering projection with the following weaker idea:

Definition I.14. A continuous surjection $f : X \to Y$ of topological spaces is a quasi-covering projection if for every connected component C of Y the restriction of f to the inverse image of C is a covering space of C.

When Y is totally disconnected (e.g., profinite) then every surjection with range Y is a quasi-covering projection, and since not all of these are covering spaces, the first notion is the more general. Since we're dealing here entirely with such totally disconnected spaces, perhaps some explanation of why surjections are to be regarded as quasi-covering projections is in order. The reason is so that the (Gleason) projective cover of a profinite space can be regarded as a universal quasi-covering projection (in the sense made precise in Theorem I.24 below). We turn now to this concept of a projective cover. The treatment here follows closely Gleason's original article, Projective Topological Spaces, which appeared in the Illinois Journal of

Mathematics, Volume 2, 1959, pages 482-489. The essential concept is the idea of an extremely disconnected space:

Definition I.15. A topological space is extremely disconnected if the closure of every open set is also open.

In a compact, Hausdorff extremely disconnected space any two distinct points p and q can be separated (let U and V be disjoint open neighborhoods of p and q, respectively. Then the closure of U is an open-closed set containing p but not q.), so such a space is profinite. It will be shown that the profinite spaces without any proper quasi-covering spaces are precisely the extremely disconnected ones (see Theorem I.19 below). To state the results efficiently, and for later use, as additional definition is called for.

Definition I.16. A continuous surjection $f : X \to Y$ of topological spaces is minimal if it carries proper closed subsets of X to proper subsets of Y.

Lemma I.17. A minimal map from a compact Hausdorff space to an extremely disconnected compact space is a homeomorphism.

Proof: Let $f : X \to Y$ be the map in question. It must be shown that f is one-one. Suppose that $a \neq b$ are elements of X. Let E and F be disjoint open neighborhoods of a and b, respectively. The complements H and K of E and F are compact, and so the complements L and M of $f(H)$ and $f(K)$ are open. L and M are disjoint since $X = H \cup K$ and f is surjective. The closure P of L is disjoint from M and since P is also open, the closure Q of M is disjoint from P. We claim that also $f(E)$ is contained in P. For let y be in $f(E)$ and let U be an open neighborhood of y. Since $E \cap f^{-1}(U) = V$ is a non-empty open subset of X, the image of its complement is not all of Y. Let $z = f(c)$ lie outside this image. Then z can't lie in the image of the complement of E. We have c in V, so that $z = f(c)$ is in $f(f^{-1}(U)) = U$. Therefore z belongs to $U \cap L$. This works for every such U, and thus y belongs to P. This establishes the claim. A similar argument shows that $f(F)$ is contained in Q. Since $f(a)$ is in $f(E)$ hence in P and $f(b)$ is in $f(F)$ hence in Q, $f(a)$ and $f(b)$ are distinct. Thus f is one-one.

<u>Lemma I.18</u>. Let $f : X \to Y$ be a continuous surjection between compact Hausdorff spaces. Then there is a closed subset Z of X such that f restricted to Z is minimal.

<u>Proof</u>: Let $\mathcal{C}$ be the set of closed subsets of X which f maps onto Y and partially order $\mathcal{C}$ by inclusion. Let $\{F_a\}$ be a descending chain of elements of $\mathcal{C}$, and let F_0 be their intersection. Let y be in Y. Then the sets $F_a \cap f^{-1}(y)$ are a descending chain of non-empty closed compact sets and hence have non-empty intersection. An element x of this intersection belongs to F_0 and goes to y,

so $f(F_0) = Y$. Thus F_0 belongs to $\mathcal{C}$. Zorn's lemma now guarantees the existence of a minimal element Z of $\mathcal{C}$ and the restriction of f to Z is a minimal map.

Theorem I.19. Let Y be a compact Hausdorff space. Every continuous surjection with compact Hausdorff domain and range Y has a continuous section if and only if Y is extremely disconnected.

Proof: Suppose Y is extremely disconnected, X is compact Hausdorff, and $f : X \to Y$ is a continuous surjection. Use Lemma I.18 to find a closed subset Z of X such that the restriction of f to Z is minimal. By Lemma I.17, Z is homeomorphic to Y . The inverse of this homeomorphism followed by the inclusion of Z into X is the desired section.

Suppose conversely that Y has the above section property. Let U be open in Y and let X be disjoint union of the closure of U and the complement of U . Map X to Y in the obvious way; this map has a continuous section s . Then the s-inverse image C of the closure of U is open and closed, contains U , and is disjoint from the complement of U . So C is the closure of U and the closure of U is open.

Theorem I.19 gives an easy way to see that the Stone-Cech compactification $\beta(D)$ of a discrete space D is extremely disconnected. $\beta(D)$ is compact and Hausdorff. If $f : X \to \beta(D)$ is a continuous surjection, construct a map $h : \beta(D) \to X$ which is a section of f over D. Then fh and the identity agree on D, and hence are equal. As another application, we have:

Proposition I.20. Let X be an extremely disconnected compact Hausdorff space, Z a closed subset of X and suppose there is a continuous $f : X \to Z$ which is the identity on Z. Then Z is extremely disconnected.

Proof: Let Y be a compact Hausdorff space and $g : Y \to Z$ a continuous surjection. Let $Y_0 = X X_Z Y = \{(x,y) \in X x Y \mid f(x) = g(y)\}$. Y_0 is compact and Hausdorff and maps onto X by projection on the first factor. This map has a section and this section followed by projection on the second factor is a section of g.

We now turn to the problem of constructing certain extremely disconnected spaces.

Theorem I.21. Let X be a compact Hausdorff space. Then there is an extremely disconnected compact Hausdorff space Y and a minimal map $f : Y \to X$.

Proof (Wilansky): Let D be X with the discrete topology and $p : \beta(D) \to X$ the map induced from the identity function. Let Y be a closed subspace of $\beta(D)$ such that p restricted to Y is minimal (use Lemma I.18 to produce Y). Let s be any (discontinuous section of p restricted to Y, and extend s to a continuous map r of $\beta(D)$ to Y. We claim that r is the identity on Y. (If it is, by Proposition I.20 Y is extremely disconnected and the theorem follows.) Suppose the claim is false, and there is a y in Y with $r(y) \neq y$. Let U and V be disjoint neighborhoods of $r(y)$ and y in Y, let $A = V \cap r^{-1}(U)$ and let B be the complement in Y of A. B is a proper closed subset of Y, as t is not in B. Since p is minimal, $p(B) \neq X$. Let x be in $X - p(B)$. If sx is in B, $x = p(sx)$ is in $p(B)$, so sx must be in A. Then $r(sx)$ is in U, so $r(sx)$ is not in V, and so also not in A. Then $r(sx)$ belongs to B. But since pr and p agree on D which is dense in $\beta(D)$, they agree, so $pr(sx) = p(sx) = x$ is in $p(B)$. This contradicts the choice of x. Thus the claim holds.

Definition I.22. Let X be a compact Hausdorff space. A Gleason (projective) cover of X is pair (Y,p) where Y is an extremely disconnected compact Hausdorff space and $p : Y \to X$ is a minimal map.

In the language of Definition I.22, Theorem I.21 asserts the existence of Gleason covers. We next consider uniqueness.

Proposition I.23. Let X be a compact Hausdorff space, (Y,p) and (Z,q) Gleason covers of X. Then there is a homeomorphism $f : Y \to Z$

such that $qf = p$.

Proof: YX_XZ is a compact Hausdorff space mapping surjectively to Y by projection on the first factor, so, since Y is extremely disconnected, has a section. This section followed by projection on the second factor gives a continuous $f : Y \to Z$ with the property that $qf = p$. If the closed subset F of Y maps onto Z by f , then $p(F) = q(f(F)) = X$, so since p is minimal $F = Y$. Thus f is a minimal map. (Since $f(Y)$ is closed in Z and maps onto X , $f(Y) = Z$.) By Lemma I.17, f is a homeomorphism.

To help motivate some later constructions, the last few results will be restated for quasi-coverings of profinite spaces.

Theorem I.24. Let X be a profinite space and (Y,p) a Gleason cover of X . Then p is a quasi-covering projection and, if $f : Z \to X$ is any quasi-covering projection, there is a continuous map $g : Y \to Z$ such that $fg = p$. These conditions plus the minimality of p determine (Y,p) up to homeomorphism over X .

Proof: Since X and Y are profinite, p is a quasi-covering projection. YX_XZ is a compact Hausdorff space mapping onto to Y , so has a section. This section, followed by projection on Z , is the desired map g . Let (W,q) be a minimal quasi-covering projection of X which maps to all quasi-covering projections as above. Then there is a continuous map $h : W \to Y$ such that $ph = q$. The minimality of p and q implies that h is a minimal map, and hence by Lemma I.17 a homeomorphism.

We next consider profinite group actions on profinite spaces. Recall that a group is <u>profinite</u> if it is an inverse limit of finite groups. This makes the group itself a topological group whose underlying topological space is profinite. A topological group whose underlying space is profinite is automatically a profinite group (the interested reader can find a proof in Cassels and Froelich, <u>Algebraic Number Theory</u>, pp. 116-127, as well as a treatment of the other facts about profinite groups used here).

A continuous action of the topological group G on the topological space X is a continuous function $G \times X \to X$, the image of (g,x) being denoted gx, such that $1x = x$ and $g(hx) = (gh)x$. The quotient X/G of X by G is the quotient space of X under the equivalence relation $\{(x,y) : x = gy \text{ for some } g \text{ in } G\}$.

<u>Lemma I.25.</u> Let the profinite group G act continuously on the profinite space X. Then X/G is profinite.

<u>Proof:</u> X/G is a continuous image of x and hence compact. Let x and y be distinct points of X/G and let a and b in X map to x and y respectively. Then the fibre over x is Ga and the fibre over b is Gb (where for S a subset of G and T a subset of X, ST is the image of $S \times T$ under the action map), and Ga and Gb are disjoint. We claim that there is an open-closed set U containing a

such that $GU \cap Gb = \phi$. If not, for every open-closed set V containing x let $F(V) = \{(g,v) \in G \times V : gv = b\}$. By hypothesis, $F(V) \neq \phi$. Since $F(V_1 \cap V_2) \subseteq F(V_1) \cap F(V_2)$, the $F(V)$ are a family of (closed) sets with the finite intersection property. Thus there is a point (g,x) belonging to all of them. Since x belongs to all V , $x = a$, and hence $Ga \cap Gb$ contains b , contrary to the hypothesis. Let U be as in the claim. Then GU is open-closed also (closed because it's the image of $G \times U$ under the action and open because it's the union of the open sets gU for all g in G). So we can assume $GU = U$. If U' is the complement of U , then also $GU' = U'$. Then the images A and A' of U and U' in X/G have U and U' as their inverse images, so A and A' are open. Since A and A' are disjoint and union up to X/G , they're also closed and they separate x and y . By Propositions I.5 and I.8, then, X/G is profinite.

Let the profinite group G act on the profinite space X . The map $X \rightarrow X/G$ need not have a continuous section, even if G is finite (see, for example, Arens and Kaplansky, Topological representations of algebras, _Trans. Amer. Math. Soc._ 63 (1948), p. 477). Since we'll need such sections later, we consider a sufficient condition for them to exist.

Definition I.26. The group G acts effectively on the space X if, for all x in X , $gx = x$ implies $g = 1$.

Lemma I.27. Let the finite discrete group G act continuously and effectively on the profinite space X. Then $X \to X/G$ has a section.

Proof: Let x be in X. Choose an open and closed neighborhood U_g of gx for each g in G such that the U_g are mutually disjoint. Let V be the intersection over G of $g^{-1}U_g$. Then V is open and closed, contains x, and $gV \cap V$ is empty if g is not the identity. Let p be the projection from X to X/G. Then $p^{-1}(pV) = \bigcup gV$ is open and closed, and hence pV is open and closed. Since p restricted to V is one-one, p has a section over $p(V)$. Thus every point of X/G has an open-closed neighborhood over which p has a section. Such neighborhoods cover X/G and hence this cover may be refined by a partition $U_1, \cdots, U_k$ (since X/G is profinite). Then over each U_i p has a section and these piece together to give a section over all of X/G.

We want to extend Lemma I.27 to treat the case where G is an arbitrary profinite group. The technique is to find sections using finite quotients of G and piece together by a 'transfinite induction' (i.e., Zorn's lemma). First we need to check that if H is a closed normal subgroup of the profinite group G which acts on the profinite space X then the profinite group G/H can be made to act on the profinite space X/H: the coset gH acting on the image of x in X is to be the image of gx. It will be left to the reader to verify that this action is well-defined, continuous, and effective if G

acts effectively on X .

Proposition I.28. Let the profinite group G act continuously and effectively on the profinite space X . Then the quotient map $p : X \to X/G$ has a section.

Proof: Consider the set $\mathcal{S}$ of pairs (T,s) where T is a closed normal subgroup of G and $s : X/G = (X/T)/(G/T) \to X/T$ is a continuous section. Define $(T,s) > (T',s')$ if $T \subset T'$ and s' is s composed with the canonical projection of X/T to X/T' . This partially orders $\mathcal{S}$. Suppose (T_a,s_a) $(a \in A)$ is an increasing chain in $\mathcal{S}$. Let $T = \bigcap T_a$. Then $X/T = \text{proj lim } X/T_a$; let s be the map of X/G to the projective limit induced by the s_a . Then (T,s) is in $\mathcal{S}$ and dominates the chain. By Zorn's lemma there is a maximal pair (N,t) in $\mathcal{S}$. Let M be an open normal subgroup of N . Then $G' = N/M$ is finite, and X/N is the orbit space of X/M under G' . G' acts effectively on X/M and hence, by Lemma I.27, there is a section $X/N \to X/M$. Since (N,t) is maximal, $N = M$. This is true for every open normal M , so $N = 1$ and $t : X/G \to X$ is a section of p .

The final topological topic to be investigated in this preliminary chapter is function spaces over profinite spaces.

Definition I.29. If X and Y are topological spaces let $C(X,Y)$ denote the space of continuous functions from X to Y with the compact open topology (i.e., a subbasic open set is the set (F,U) of all functions carrying the compact set F into the open set U).

Proposition I.30. Let X and Z be profinite spaces, Y a discrete space.

a) $C(X,Y) = \text{dir lim } C(P,Y)$ where P ranges over the partitions of X .

b) $C(X,Y)$ is discrete and the adjoint map $C(X,C(Z,Y)) \to C(X\text{x}Z,Y)$ is a bijection.

c) Let W be a closed subset of X . Then the restriction $C(X,Y) \to C(W,Y)$ is onto.

Proof: a) For every partition P of X there is a (continuous) surjection $X \to P$ which assigns to x in X the element of P to which x belongs. This induces an injection $h_P : C(P,Y) \to C(X,Y)$. These maps are compatible with refinements of partitions and induce an injection $h : \text{dir lim } C(P,Y) \to C(X,Y)$. Let $f : X \to Y$ be any continuous function. Since $f(X)$ is a compact subset of the discrete space Y , it is finite; say $f(X) = \{a_1,\cdots,a_n\}$ where the a_i are all distinct. Let $U_i = f^{-1}(a_i)$. Then $P = \{U_1,U_n\}$ is a partition of X . If g is the function on P defined by $g(U_i) = a_i$ then $f = h_p(g)$ and hence f is in the image of h . Thus h is a bijection.

b) To see that $C(X,Y)$ is discrete, let f be in $C(X,Y)$. Then if $f(X) = \{a_1,\cdots,a_n\}$, $\bigcap_i (f^{-1}(a_i),a_i) = \{f\}$ is open. The second assertion is a standard fact (see, for example, Dugundji, Topology, Allyn and Bacon, Boston, 1966, p. 261).

c) Let $f : W \to Y$ be continuous, let $f(W) = \{a_1, \cdots, a_n\}$ where the a_i are all distinct, let $U_i = f^{-1}(a_i)$ and let $P = \{U_1, U_n\}$. P is a partition of W . We'll produce a partition P' of X such that $P' \cap W = P$. U_1 and $U = U_2 \cup \cdots \cup U_n$ are disjoint closed subsets of the profinite space X . There are, by Lemma I.4, open-closed disjoint subsets V_1 and V of X containing U_1 and U respectively whose union is X . Split V into two open-closed parts, one containing U_2 and the other the rest as above and continue this process to produce a partition $P' = \{V_1, \cdots, V_n\}$ of X such that for each i , U_i is contained in V_i . Define $g : X \to Y$ by $g(x) = a_i$ if x is in V_i . Then f is the image of g under $C(X,Y) \to C(W,Y)$.

Now suppose R is a commutative ring and M is an R-module (R-algebra). Consider M as a discrete topological space. Then for any topological space X , $C(X,M)$ is an R-module (R-algebra) under pointwise operations. We'll need to know how these constructions behave under tensor products.

Proposition I.31. Let X be a profinite space, R a commutative ring and M , N R-modules (R-algebras) considered as discrete spaces. Then the map

$$C(X,M)\otimes_R N \to C(X,M\otimes_R N)$$

defined by $(f \otimes n)(x) = f(x) \otimes n$ is an R-module (R-algebra) isomorphism

Proof: If X is finite, the assertion is the distributive law of tensor products over direct sums. In general, we have the chain of isomorphisms (as P ranges over the partitions of X) : $C(X,M)\otimes_R N =$ $\text{dir lim}\, C(P,M)\otimes_R N = \text{dir lim}\, C(P,M\otimes_R N) = C(X,M\otimes N)$, using Proposition I.30 (a).

Bibliographic Note on Chapter I

The material on profinite and totally disconnected spaces comes from Hurewicz and Wallman, *Dimension Theory*, Princeton, 1948, and Wilansky, *Topology for Analysis*, Ginn, Waltham, Mass., 1970. Some of the material on extremely disconnected spaces is from Wilansky; the rest is from Gleason's article quoted above. The remainder of the chapter is largely folklore.

II. The Boolean Spectrum.

The presence of idempotents significantly complicates the Galois theory of separable, projective algebras over commutative rings - as these notes indicate. Of course Galois theory is not unique in this respect. A similar thing happens in the theory of commutative rings which are regular in the sense of von Neumann: every element of such a ring is a unit times an idempotent, so, except for idempotents, the theory of these rings should be the same as the theory of fields. What is needed is a systematic way to discard and recover idempotents. In the easiest case of commutative regular rings, where the only unit is 1 (i.e., a Boolean ring), one can think of the Stone Representation Theorem, which displays every Boolean ring as the ring of all continuous Z/2Z-valued functions on its maximal ideal space, as such a discard-recovery technique.

In his study, Modules over Commutative Regular Rings (Amer. Math. Soc. Memoirs No. 70, 1967), R. S. Pierce discovered a technique which works for all regular, in fact all commutative, rings. The idea is to display the ring in question as the ring of all global sections of a sheaf of rings such that each stalk has no idempotents except zero and one. Of course, the usual representation of a commutative ring as sections of its sheaf of local rings over the prime spectrum also has the same properties. Since Pierce only wants to discard idempotents instead of prime ideals, however, he can work with the minimal sheaf to do this, and the representation he produces has the advantage that the maps from global sections to stalks are surjective and the base space of the sheaf is profinite. The construction of the representation mimics that of the construction of the usual sheaf of local rings, except to use instead the prime ideals and prime spectrum of the (Boolean) ring of idempotents of the ring which is being represented.

In fact, Pierce was also interested in non-commutative rings, so his construction is a bit more general than has been indicated here. If commutative rings alone are to be studied, there is an easier construction, which Villamayor and Zelinsky pointed out in their paper, "Galois theory with infinitely many idempotents" (*Nagoya Math. J.* 35(1969)): take the ring R and represent it in the usual way as the sections of the sheaf $\mathcal{R}$ of local rings on the prime spectrum $\text{Spec}(R)$ of R. Let $q : \text{Spec}(R) \to \text{Comp}(\text{Spec}(R))$ be the map which identifies connected components to points. Then Pierce's sheaf is the direct image $(\text{Comp}(\text{Spec}(R)), q_*\mathcal{R})$.

Villamayor and Zelinsky went on to show further that the basic properties of the sheaf which figure in Galois theory, as opposed to representation theory, could easily be stated and proved without the language of sheaves. This approach has some obvious expository advantages, and will be adopted here. We begin with the definition of the Boolean spectrum and a look at its topology.

Definition II.1. Let R be a commutative ring. $\text{Spec}(R)$, as usual, denotes the set of prime ideals of R endowed with the topology which makes, for every ideal I of R, the set $V(I) = \{p \in \text{Spec}(R) \mid p \geq I\}$ a closed set. Then $X(R)$ denotes the space $\text{Comp}(\text{Spec}(R))$ (see I.11) above. We'll refer to $X(R)$ as the <u>Boolean spectrum</u> of R, for reasons which will become clear shortly (see II.9) below.

If $f : R \to S$ is a ring homomorphism there is an induced continuous function $\text{Spec}(S) \to \text{Spec}(R)$. The continuous map $\text{Spec}(S) \to \text{Spec}(R) \to X(R)$ then collapses connected components to points and so induces a continuous function $X(f) : X(S) \to X(R)$. It's easy to verify that this makes $X(\cdot)$ into a functor.

<u>Lemma II.2</u>. Every open-closed subset of $\text{Spec}(R)$ is of the form $V(Re)$ for some idempotent e of R. In particular, $\text{Spec}(R)$ is connected if and only if R has no idempotents except zero and one.

<u>Proof</u>: Let $V(I)$ be an open-closed subset of $\text{Spec}(R)$. Then there is an ideal J of R such that $V(I) \cup V(J) = \text{Spec}(R)$ and $V(I) \cap V(J)$ is empty. The former says that IJ is nilpotent and the latter that $I + J = R$. Write $1 = i + j$ with i in I, j in J and suppose $(ij)^n = 0$. Then $1 = 1^{2n} = i^n r + j^n s$ where r,s are in R. Replace i by $i^n r$, and j by $j^n s$. Then we can assume $1 = i + j$ with i in I, j in J and $ij = 0$. So i is an idempotent. If a prime ideal contains I it contains i, and if it contains i it can't contain j nor J, so $V(I) = V(Ri)$.

Conversely, if e is an idempotent of R then $V(Re) \cap V(R(1-e))$ is empty (no proper prime contains both e and $1 - e$) while $V(Re) \cup V(R(1-e)) = \text{Spec}(R)$ (since $e(1 - e) = 0$, every prime contains either e or $1 - e$) so $V(Re)$ is open-closed.

Because of II.2, if R has no idempotents except 0 and 1 we call R _connected_.

Proposition II.3. Two primes P and Q of R belong to the same connected component of Spec (R) if and only if they contain the same idempotents.

Proof: First, if e is an idempotent of R belonging to P but not to Q, then $V(Re)$ is an open-closed set containing P but not Q, so P and Q are in different components. Now suppose every idempotent of P belongs to Q and let I be the ideal of R generated by the idempotents of P. Then P and Q belong to $V(I)$. $V(I)$ is the image of Spec (R/I) under the continuous map Spec $(R/I) \to$ Spec (R) induced from the projection $R \to R/I$. In II.21 below R/I is shown to be connected. So $V(I)$ is a connected set and it follows that P and Q belong to the same component.

Proposition II.3 and the first part of its proof show that points of Spec (R) in different components can be separated by open-closed sets. Since Spec (R) is compact this means, using I.4, that distinct components of Spec (R) can be separated. Thus, using I.13, we have:

Corollary II.4. $X(R)$ is a profinite space.

Because of Proposition II.3, one would like to know which sets of idempotents can occur as the set of all idempotents in a prime ideal.

This is answered by the following definition and proposition.

Definition II.5. A set of idempotents E of the commutative ring R is a maximal Boolean ideal if:

i) For every idempotent e of R either $e \in E$ or $1 - e \in E$, but not both;

ii) If e and f are idempotents of R then $ef \in E$ if and only if $e \in E$ or $f \in E$.

Proposition II.6. A set E of idempotents of R is the set of all idempotents in some prime ideal if and only if E is a maximal Boolean ideal.

Proof: Let P be a prime ideal of R and E the set of all idempotents in it. Then if e is an idempotent, since $e(1 - e) = 0$ is in P, either e or $1 - e$ is in P; but not both since $1 = e + (1 - e)$ is not in P. If e,f are idempotents of R then $ef \in P$ if and only if $e \in P$ or $f \in P$. Since ef is an idempotent, this means $ef \in E$ if and only if $e \in E$ or $f \in E$. So E is a maximal Boolean ideal. Now suppose we're given a maximal Boolean ideal E. Suppose $RE = R$. Then $1 = r_1e_1 + \cdots + r_ke_k$ with each e_i in E. Let $f_i = e_i(1-e_1)(1-e_2)\cdots(1-e_{i-1})$. Then the ideal generated by the f_i's is the same as the ideal generated by the e_i's, so $1 = s_1f_1 + \cdots + s_kf_k$, for some k. We note that each f_i is in E and that $f_if_j = 0$ if $i \neq j$. Multiplying the above equation by each f_i in turn, we see that $s_if_i = f_i$. So $1 = f_1 + \cdots + f_k$. Then $0 = (1-f_1)(1-f_2)\cdots(1-f_k)$. But 0 is in E since $0 = 0\cdot f_1$ while no $1 - f_i$ belongs to E. This contradicts part ii) of Definition II.5. So RE is contained in some maximal ideal M of R. If e is an idempotent of M and if $1 - e$ belonged to E then 1 would belong to M, so $1 - e$ can't be in E and e must be. Thus E contains exactly the idempotents of M.

Finally, we'll explain the terminology of Definition II.5.

Definition II.7. $B(R)$ denotes the set of idempotents in the commutative ring R.

We make $B(R)$ into a commutative ring with identity by defining products to be the same as products in R and sums as follows: if $e, f \in B(R)$, $e \oplus f = e + f - ef$ (operations in R). (The interested reader is invited to verify that these operations do indeed make $B(R)$ a commutative ring.)

We leave the proofs of the following facts as exercises as they will not be used later:

Proposition II.8. E is a maximal Boolean ideal of R if and only if E is a maximal ideal of $B(R)$; moreover, all prime ideals are maximal.

Proposition II.9. The function which associates to each component of Spec (R) the set of idempotents of any element of that component defines a homeomorphism of $X(R)$ to Spec $(B(R))$.

Next, we'll take a look at the topology of $X(R)$. We saw above (II.4) that $X(R)$ is profinite. By I.8, this means that $X(R)$ has a basis of open-closed sets. Since an open-closed subset of Spec (R) contains every connected component it meets, such a set is the inverse image of its projection down into $X(R)$, and hence its projection into $X(R)$ is an open-closed subset. Conversely, an open-closed subset of $X(R)$ has open-closed inverse image in Spec (R) and is the projection of its inverse image. These facts, plus the description of open-closed subsets of Spec (R) recorded above in Lemma II.2 allows us to describe the topology of $X(R)$. First a definition:

Definition II.10. Let e be an idempotent of R . Then let $N_R(e) = \{x \in X(R) : x \subset V(R(1-e))\}$. (If no confusion arises, we'll usually drop the subscript R .)

Proposition II.11. The sets $N(e)$, as e ranges over the idempotents of R , form a basis of open-closed sets for the topology of $X(R)$; moreover, every open-closed subset of $X(R)$ is of the form $N(e)$ for a suitable idempotent e of R .

Proof: We need only observe that $N(e)$ is the projection into $X(R)$ of the open-closed subset $V(R(1-e))$ of Spec (R) , and then apply the above remarks.

If the component x belongs to $N(e)$ and P is any element of x , then $1 - e$ belongs to P . Conversely, if P belongs to the component x and $1 - e$ belongs to P, then $1 - e$ belongs to every Q in x by II.3, so x belongs to $N(e)$.

We note the following set operations on the basis of $X(R)$:

Proposition II.12. Let e and f be idempotents of $X(R)$. Then:

i) $N(e) \cap N(f) = N(ef)$

ii) $N(e) \cup N(f) = N(e + f - ef)$

iii) $X(R) - N(e) = N(1 - e)$

iv) $N(0) = \phi$

v) $N(1) = X(R)$

Proof: We'll prove ii), leaving the remainder as exercises. If x belongs to $N(e) \cup N(f)$ and P belongs to x, then either $1 - e$ belongs to P or $1 - f$ belongs to P, so $(1 - e)(1 - f) = 1 - (e + f - ef)$ belongs to P and hence x belongs to $N(e + f - ef)$. If conversely x belongs to $N(e + f - ef)$ and P belongs to x, then $(1 - e)(1 - f)$ belongs to P, so either $1 - e$ or $1 - f$ is in P. In the former case x is in $N(e)$ and in the latter x is in $N(f)$.

Because of II.12 ii) we will denote the idempotent $e + f - ef$ by $e \cup f$.

Corollary II.13. $N(e) = N(f)$ if and only if $e = f$.

Proof: Suppose $N(e) = N(f)$. Then $N((1 - e)f) = N(1 - e) \cap N(e) = \phi$ by II.12 i) and iii). So $1 - (1 - e)f$ is an idempotent belonging to no prime ideals, so it must be 1. Then $(1 - e)f = 0$ so $f = ef$.

Similarly $(1 - f)e = 0$ so $e = fe$ and hence $e = f$.

As a consequence, if $N(e) \cap N(f) = \phi$, $ef = 0$ so $e \cup f = e + f$.

Our next task is to associate to each component x of Spec (R) and each R-module M an R-module M_x. (M_x is the stalk at x of the sheaf over $X(R)$ corresponding to the module M is Pierce's representation theory, although we won't use that interpretation here.)

Definition II.14. Let M be an R-module and x an element of $X(R)$. Let $I(x)$ denote the ideal generated by the set of all idempotents in any prime ideal belonging to x. Then let $R_x = R/I(x)$ and $M_x = M \otimes_R R_x$ $(= M/I(x)M)$. If a is an element of M, a_x denotes the image of a in M_x. If $f : M \to N$ is an R-module homomorphism, let $f_x : M_x \to N_x$ be $f \otimes R_x$.

If e is in idempotent of R, x is in $N(e)$ if and only if $e_x = 1_x$.

We observe that there is a canonical surjection $M \to M_x$ for each x in $X(R)$. We're now going to show that the M_x's act like the stalks of a sheaf. First a lemma:

Lemma II.15. Let M be an R-module and I an ideal of R generated by idempotents. Then x in M belongs to IM if and only if $x = ex$ for some idempotent e in I.

Proof: If $x = ex$ then x belongs to IM. So suppose x belongs to

IM , i.e., $x = e_1x_1 + \cdots + e_kx_k$ where e_i is an idempotent in I and x_i belongs to M . Let $e = e_1 \cup e_2 \cup \cdots \cup e_k$. Then for each i , $ee_i = e_i$. It follows that $ex = x$.

Proposition II.16. Let a and b belong to the R-module M , let x be in X(R) and suppose $a_x = b_x$. Then there is a neighborhood N(e) of x in X(R) such that $a_y = b_y$ in M_y for all y in N(e) .

Proof: If $a_x = b_x$ then a - b belongs to I(x)M . By Lemma II.15, there is an idempotent e in I(x) such that a - b = e(a - b) . The idempotent e then belongs to any prime ideal in the component x , so that x is in N(1 - e) . If y belongs to N(1 - e) then e belongs to I(y) , so a - b belongs to I(y)M and $a_y = b_y$.

We'll often use Proposition II.16 in the following form: If a and b in the R-module M are such that $a_x = b_x$ there is an idempotent e of R with $e_x = 1_x$ and ae = be . (The idempotent e that we're referring to in this formulation is the idempotent 1 - e produced in the proof of II.16.)

Proposition II.17. Let a and b belong to the R-module M and suppose that for all x in X(R) that $a_x = b_x$. Then a = b .

Proof: For each x in X(R) there is an idempotent e(x) such that x is in N(e(x)) and e(x)a = e(x)b by Proposition II.16. The neighborhoods N(e(x)) cover the profinite space X(R) so we can extract a finite disjoint subcover $N(e_1),\ldots,N(e_k)$. Then since

$N(e_1) \cup \cdots \cup N(e_k) = X(R) = N(1)$, we have that $e = e_1 + \cdots + e_k = 1$ (using Proposition II.12 and Corollary II.13). Since $ee_i = e_i$ for all i and $e_i a = e_i b$ for each i we have that $a = \Sigma e_i a = \Sigma e_i b = b$.

Proposition II.18. For each x in $X(R)$, R_x is a flat R-module.

Proof: Let $M \to N$ be an injection of R-modules. We want to show that $M_x \to N_x$ is also an injection, i.e., that $M \cap I(x)N = I(x)M$. But if m in M belongs to $I(x)N$ then $m = em$ for some e in $I(x)$, by Lemma II.15, so m belongs to $I(x)M$. The reverse inclusion is trivial and the proposition follows.

The reader has probably noticed the possible ambiguity in the notation f_x given in Definition II.14: if $f : M \to N$ is an R-module homomorphism and x in $X(R)$ then f_x denotes both the homomorphism $f \otimes R_x$ and the image of f in the R-module $\mathrm{Hom}_R\,(M,N)_x$. There is of course a (natural) transformation

$$(*) \qquad \mathrm{Hom}_R\,(M,N)_x \to \mathrm{Hom}_{R_x}\,(M_x,N_x)$$

which sends f_x to $f \otimes R_x$, but it need not be an isomorphism in general, so the ambiguity of notation is genuine. We have, however, the following:

Lemma II.19. If M is finitely generated, (*) is a monomorphism. If M is finitely presented, (*) is an isomorphism.

Proof: Regard both the source and target of (*) as functors of M ;

note that both functors are additive and left exact (we use II.18 here). When $M = R$, (*) is an isomorphism. Both assertions of the lemma now follow by applying the functors to exact sequences $R^{(n)} \to M \to 0$ and $R^{(n)} \to R^{(m)} \to M \to 0$.

We'll use II.16 through II.19 to lift data defined over various R_x's back to similar data over R. As a first application of this technique we have:

Proposition II.20. Let S be an R-algebra x in $X(R)$, and e_0 an idempotent of S_x. Then there is an idempotent e of S such that $e_x = e_0$.

Proof: Choose f in S such that $f_x = e_0$. Then since $f_x^2 = f_x$, there is by II.16 an idempotent e of R with x in $N(e)$ such that $ef^2 = ef$, i.e., $(ef)^2 = ef$. Since x is in $N(e)$, $e_x = 1_x$ so $(ef)_x = e_0$.

As a consequence of II.20 we see that the R_x's are connected:

Corollary II.21. Let x be in $X(R)$. Then R_x has no idempotents except zero and one.

Proof: Let e_0 be an idempotent of R_x. Choose, by II.20, an idempotent e of R such that $e_x = e_0$. Then by II.6 either $e \in I(x)$ or $1 - e \in I(x)$. In the first case $e_x = 0$ and the second $e_x = 1$.

Now we lift finitely presented modules.

Lemma II.22. Let x be in $X(R)$ and M_0 a finitely presented R_x-module. Then there is a finitely presented R-module M such that $M_x = M_0$.

Proof: Let $T_0 : (R_x)^{(n)} \to (R_x)^{(m)}$ be an $m\times n$ matrix such that $\text{Coker}(T_0) = M_0$. Choose $T : R^{(n)} \to R^{(m)}$ such that $T_x = T_0$. Let $M = \text{Coker}(T)$. Then $M_x = M_0$.

Note that the crucial property in the proof of II.21 is the surjectivity of $R \to R_x$. In order to lift projective modules, which we now do, we'll also use the following interpretation of II.16 and II.19: if equations (diagrams) of maps between finitely presented modules hold (commute) at x, they do so in a neighborhood of x.

Proposition II.23. Let S be an R-algebra, finitely generated and projective as an R-module. Let N_0 be a finitely generated projective S_x-module (x in $X(R)$). Then there is a finitely generated projective S-module N such that $N_x = N_0$.

Proof: N_0 is finitely presented as an R_x-module. So by II.22 there is a finitely presented R-module M such that $M_x = N_0$ (as R_x-modules). To say that M has an S-module structure is to give a map $S\otimes_R M \to M$ satisfying an associative law, i.e., making certain diagrams (whose vertices are finitely presented R-modules) commute. Since $S\otimes_R M$ is finitely presented we can, by II.19, choose an R-module map k lifting the S_x-module structure on M_x . Then the diagram for the associativity of k commutes at x and so, by II.16 in a neighborhood of x , i.e., there is an idempotent e of R with $e_x = 1_x$ such that ke makes Me an Se-module. Replace M by $S(1 - e) \oplus Me$. Then M is an S-module and $M_x = N_0$ as S_x-modules. Since M is finitely generated over S there is for some n an S-module epimorphism $g : S^{(n)} \to M$. Let h_0 be the right S_x-module inverse to g_x (which exists since $M_x = N_0$ is S_x-projective). Let $h : M \to S^n$ be an R-module homomorphism so that $h_x = h_0$. To say that h is an S-module homomorphism is again an assertion that a diagram with finitely presented vertices commutes. To say that h is a right inverse to g is to assert that an equation between homomorphisms of finitely presented modules holds. The diagram commutes and the equation holds at x . So by II.16 similar things happen in a neighborhood of x ; that is, there is an idempotent f of R with $f_x = 1_x$ such that hf is an Sf-module homomorphism and a right inverse to gf , so Mf is Sf-projective. Then $N = S(1 - f) \oplus Mf$ is a finitely generated projective S-module such that $N_x = N_0$.

Now we need also to lift algebras.

<u>Proposition II.24</u>. Let x be in $X(R)$ and S_0 an R_x-algebra, finitel presented as an R-module. Then there is a finitely presented R-algebra S such that $S_x = S_0$. If S_0 is a projective R_x-module, S can be

taken projective. If S_0 is a separable R_x-algebra, S can be taken separable.

Proof: By II.22 there is a finitely presented R-module T such that $T_x = S_0$. To say that T is an R-algebra is to give maps $T\otimes_R T \to T$ and $R \to T$ (multiplication and identity) which make certain diagrams commute. Since R and $T\otimes_R T$ are finitely presented we can choose maps lifting the multiplication and identity of S_0 . So there is an idempotent e with $e_x = 1_x$ such that $S = R(1 - e) \oplus Te$ is an R-algebra with $S_x = S_0$. (If S_0 is a finitely generated projective R-module then by II.23 we can take T to be projective and hence S is also projective.) S is separable if the multiplication map $m : S\otimes_R S \to S$ has a right $S\otimes_R S$-module inverse. Exactly as in the proof of II.23, if such an inverse exists at x (i.e., if S_0 is separable) it exists on a neighborhood: i.e., there is an idempotent f of R such that $f_x = 1_x$ and Sf is Rf-separable. Thus replacing S by $R(1 - f) \oplus Sf$ we have S also separable (and projective if our originial S was).

Finally, we will have to lift algebra homomorphisms:

Lemma II.25. Let S and T be R-algebras with S finitely presented as an R-module and let x be in X(R) . Suppose that there is an R-algebra homomorphism $h_0 : S_x \to T_x$. Then there is an idempotent e of R with $e_x = 1_x$ and an Re-algebra homomorphism $h : Se \to Te$ such that $h_x = h_0$.

Proof: By II.19 there is an R-module homomorphism $g : S \to T$ such that $g_x = h_0$. Let $a^1,\ldots,a^n$ generate S as an R-algebra. Then since $g(a^i a^j)_x = g_x(a^i_x a^j_x) = h_0(a^i_x a^j_x) = h_0(a^i_x)h_0(a^j_x) = g(a^i)g(a^j)_x$ there is by II.16 an idempotent e with $e_x = 1_x$ such that $ge(a^k e a^j e) = ge(a^i e)ge(a^j e)$, so $ge : Se \to Te$ is an Re-algebra homomorphism.

Proposition II.26. Let S and T be R-algebras, finitely generated and projective as R-modules. Suppose for some x in $X(R)$ that S_x and T_x are isomorphic. Then there is an idempotent e of R with $e_x = 1_x$ such that Se is isomorphic to Te as Re-algebras.

Proof: Let $h_0 : S_x \to T_x$ be an R_x-algebra isomorphism. By Lemma II.25 there is an e' with $e'_x = 1_x$ and an Re'-algebra homomorphism $h : Se' \to Te'$ such that $h_x = h_0$; replace R,S,T by Re',Se',Te' . Let $b^1,\ldots,b^n$ generate T over R . Since h_0 is onto there are $a^1,\ldots,a^n$ in S such that $h(a^i)_x = h_0(a^i_x) = b^i_x$. So there is an idempotent e with $e_x = 1_x$ such that $he(a^i e) = b^i e$; this means that $he : Se \to Te$ is onto. Since Te is projective the kernel N of he is a direct summand of Se and hence finitely generated, say by $c^1,\ldots,c^k$, over R . Since R_x is R-flat (II.18) N_x is the kernel of h_0 , hence $c^i_x = 0$ for all i . Thus there is an idempotent f with $f_x = 1_x$ such that the $c^i f = 0$. Replacing e by fe we have that $hf : Sf \to Tf$ is an isomorphism.

The above result is just one of many which assert "if a property holds at one point of $X(R)$ it holds in a neighborhood of that point" - in the context of II.26, if e is the idempotent produced in the proposition then x is in $N(e)$ and if y is any point of $N(e)$, S_y and T_y are isomorphic as R_y-algebras. Now suppose we know a certain property holds at every point of $X(R)$. Then $X(R)$ is covered by neighborhoods on which the property holds. Since $X(R)$ is profinite we can refine this open cover by a partition. The open sets in this partition do not overlap, so we can paste together without worrying about consistency conditions. This is a powerful technique which will be used often, and we give a first example of it now.

Proposition II.27. Let S and T be R-algebras, finitely generated and projective as R-modules. Suppose that for each x in $X(R)$ S_x and T_x are isomorphic R_x-algebras. Then S and T are R-algebra isomorphic.

Proof: By II.26 there is for each x in $X(R)$ an idempotent $e(x)$ such that $x \in N(e(x))$ and $Se(x)$ and $Te(x)$ are $Re(x)$-algebra isomorphic. $\{N(e(x)) : x \varepsilon X(R)\}$ is an open cover of $X(R)$. Let $\{U_1,\ldots,U_n\}$ be a partition refining this open cover. By II.11, each $U_i = N(e_i)$ for suitable idempotents e_i . Then by II.12 and II.13 since we have a partition $e_i e_j = 0$ if $i \neq j$ and $e_1 + \cdots + e_n = 1$. If $N(e_i) \subseteq N(e(x))$ then $N(e_i) \cap N(e(x)) = N(e_i)$ so $e_i e(x) = e_i$ by II.12 and II.13, so Se_i and Te_i are isomorphic Re_i-algebras, say by the isomorphism h_i . Then $h = h_1 \oplus \cdots \oplus h_n : S = Se_1 \oplus \cdots + Se_n \to Te_1 \oplus \cdots \oplus Te_n = T$ is the desired R-algebra isomorphism.

The remainder of this chapter will be devoted to some examples and illustrations of the Boolean spectrum. If X is a topological space and R a topological group or ring then $C(X,R)$ denotes the set of all continuous R-valued functions on X made into a group or ring using pointwise operations (e.g., $(f + g)(x) = f(x) + g(x)$). We want to compute the Boolean spectra of such rings, at least in some special cases.

Suppose X is profinite and R is a discrete ring with no idempotents except zero and one. Let $S = C(X,R)$. We are going to show that X and $X(S)$ are homeomorphic. First, note that every function f in S is locally constant. Now let x be in X and consider the map $S \to R$ given by evaluation at x . The induced map $\text{Spec}(R) \to \text{Spec}(S)$ has image $V(I)$, where I is the kernel of the evaluation. If $f \in I$, $f(x) = 0$ so there is an open-closed neighborhood U of x in X such that $f(y) = 0$ for all y in U . Let e in S be the function whose value on U is zero and whose value elsewhere is one. Then: $fe = f$, $e^2 = e$ and $e \in I$. It follows that I is generated by idempotents so that $V(I)$ is an intersection of the open-closed sets $V(Se)$ for e an idempotent in I . Since $V(I)$ is the image of the connected space $\text{Spec}(R)$ it's also connected, and so $V(I)$ is actually a component of $\text{Spec}(S)$; call it $\phi(x)$. We've defined a map $\phi : X \to X(S)$. Now let e be an idempotent of S . Then $\phi(x)$ is in $N(e)$ if and only if there is a prime ideal P of S containing $1 - e$ with P in $\phi(x)$. This means that P contains I , so that $e \equiv 1(I)$, i.e., $e(x) = 1$. Thus

$\phi^{-1}(N(e)) = \{x \varepsilon X : e(x)=1\}$. Since e is an idempotent the only values it takes are 0 and 1 , so the set on the right is open. So ϕ is continuous. If x,y are in X , $x \neq y$, there is an open-closed set containing x but not y . The characteristic function of this open-closed set is an idempotent e of S such that $e(x) = 1$ and $e(y) \neq 1$. This means that $\phi(x) \varepsilon N(e)$ and $\phi(y) \notin N(e)$ so that $\phi(x) \neq \phi(y)$. So ϕ is an injection. Since both X and $X(S)$ are compact Hausdorff $\phi(X)$ is closed. Then there is an open-closed subset of its complement of the form $N(e)$. If $e(x) = 1$ for any x , $\phi(x)$ would be in $N(e)$. So $e(x) = 0$ for all x , i.e., $e = 0$ and $N(e)$ is empty. So ϕ is onto, and hence is a homeomorphism from X to $X(S)$.

Actually, we've shown a bit more. The ideal I we produced above was generated by all idempotents e such that $e(x) = 0$, i.e., all e such that $\phi(x) \varepsilon N(1 - e)$. But these are precisely the idempotents e which belong to $I(x)$. So I and $I(x)$ are same ideals, and we have a commutative diagram

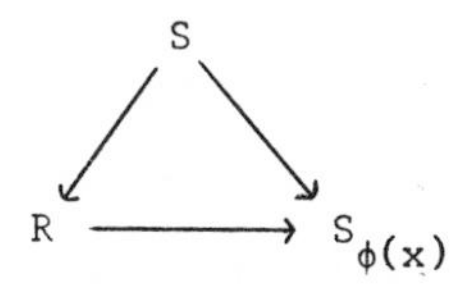

where the left map is evaluation, the right map the canonical one and the bottom one induced by the map $R \to S$.

One consequence of the above discussion is that every profinite space appears as a Boolean spectrum.

As an example, let X be the convergent sequence $\{1,\frac{1}{2},\frac{1}{3},\ldots,0\}$ (X is a profinite space), and let $\mathbb{R}$ and $\mathbb{C}$ denote the real and complex numbers, both with the discrete topology. Then let $A = C(X,\mathbb{R})$ $B = C(X,\mathbb{C})$ and $T = \{f \varepsilon B : f(0) \varepsilon \mathbb{R}\}$. By what was shown above, we can (and do) identify $X(A)$ and $X(B)$ with X. Since every idempotent of T lies in A we can also identify $X(T)$ with $X(A)$ and hence with X. Then we have that if $x \neq 0$, $T_x = \mathbb{C}$ and $T_0 = \mathbb{R}$: for any x in X, $A_x \subseteq T_x \subseteq B_x$, i.e., $\mathbb{R} \subseteq T_x \subseteq \mathbb{C}$. If $x \neq 0$, we can find f in B such that $f(x) = \sqrt{-1}$ and $f(0) = 0$. Then $f \varepsilon T$ and $f_x = f(x) = \sqrt{-1}$ is in T_x. If $f \varepsilon T$ then f_0 in T_0 is the same as f_0 in B_0 which is $f(0)$, so $T_0 \subseteq \mathbb{R}$. This example, and some easy variations on it, will be useful later.

Sometimes it's convenient to deal with rings which are products of rings of functions. As will be shown shortly (II.29), such rings arise naturally in the study of separable algebras over rings of continuous functions.

Definition II.28. R is a weakly uniform ring if there are profinite spaces $X_1,\ldots,X_n$ and discrete rings $A_1,\ldots,A_n$ with no non-trivial idempotents such that $R = C(X_1,A_1) \times \cdots \times C(X_n,A_n)$. (The terminology is due to DeMeyer, Separable polynomials over commutative rings, Rocky Mt. Journal 2, (1972); "weakly" is used so that "uniform" can be reserved for rings of functions.)

Proposition II.29. Let R be a weakly uniform ring and let S be a separable R-algebra, finitely generated and projective as an R-module. Then S is a weakly uniform ring.

<u>Proof</u>: Because we want to help the reader become familiar with some of the techniques involved, the proof will be given in detail. $R = C(X_1,A_1) \times \cdots \times C(X_n,A_n)$ where the X_i are profinite, the A_i discrete, and each A_i is connected. Let e_i be the idempotent of R such that $Re_i = C(X_i,A_i)$. Suppose we can show that Se_i is weakly uniform. Then $S = Se_1 \times \cdots \times Se_n$ will be weakly uniform. So we may assume $R = C(X,A)$. Let x be in $X(R) = X$. Then S_x is separable over $R_x = A$ and finitely generated and projective as an R_x=A-module. Let $T = C(X,S_x)$. By I.31, $T = C(X,A)\otimes_A S_x$ so T is a separable R-algebra, finitely generated and projective as an R-module. Also $T_x = S_x$ (here T_x means $T\otimes_R R_x$; since $X(T) = X(R) = X$, this is the same as T_x where x is in $X(T)$). Thus T_x and S_x are isomorphic. So by Proposition II.26 there is an idempotent e of R with x in $N(e)$ such that Se and Te are isomorphic Re-algebras. Then $Te = C(N(e),S_x)$. The e we constructed depends on x, but since x belongs to $N(e)$ we have $X(R)$ covered by such neighborhoods. We can refine that cover to a partition $\{N(e_1),\ldots,N(e_n)\}$ of $X(R)$ such that i) $e_ie_j = 0$ if $i \neq j$ and $e_1 + \cdots + e_n = 1$; and ii) $Te_i = C(N(e_i),S_i)$ (where S_i is some separable, projective A-algebra) is Re_i-algebra isomorphic to Se_i. Thus $S = Se_1 \times \cdots \times Se_n = C(N(e_1),S_1) \times \cdots \times C(N(e_n),S_n)$ is weakly uniform.

Bibliographic note on Chapter II:

We've followed here the exposition of the Boolean spectrum given by Villamayor and Zelinsky in the paper "Galois theory with infinitely many idempotents" (Nagoya J. 35(1969)); as mentioned above the theory is originally due to Pierce and appears in his *Modules over Commutative Regular Rings* (Amer. Math. Soc. Memoirs 70, 1967).

The lifting constructions (II.19 through II.25) are from "Pierce's representation and separable algebras" (Ill. J. Math. 15(1971)) and the interested reader can find further details and applications of the techniques there. II.29 appears as Proposition 2.16 in "The separable closure of some commutative rings" (Trans. Amer. Math. Soc. 170(1972)).

III. Galois Theory over a Connected Base.

In this Chapter we begin our study of the Galois theory of separable algebras, starting with the simplest case: where the base ring has no idempotents except zero and one. As we'll see in subsequent chapters, the general theory reduces to and depends on this special case.

Also, beginning with this Chapter the reader will be assumed to be familiar with most of the basic properties of commutative separable algebras. As these properties are used for the first time the appropriate reference to DeMeyer and Ingraham will be given (recall we are indicating that text by DI) but from then on the properties are considered known.

The following definitions will be important:

Definition III.1. Let R be a commutative ring and S an R-algebra. S is an extension of R if it's a faithful R-algebra. S is a strongly separable extension of R if S is an extension of R which is a separable R-algebra and a finitely generated projective R-module. S is a locally strongly separable extension of R if S is a direct limit of strongly separable subextensions (i.e., if any finite subset of S belongs to a subalgebra which is a strongly separable extension of R).

Definition III.2. Let R be a commutative ring with no idempotents except zero and one and let S be an extension of R which also has no non-trivial idempotents. S is a *Galois* extension of R if S is a strongly separable R-extension and if every element of S not in R is moved by some R-algebra automorphism of S . S is an *infinite Galois* extension of R if S is a locally strongly separable R-extension such that the fixed ring of the R-automorphisms of S is again R .

The reader should be warned that the terminology of Definition III.2 is not standard. Some authors (e.g., [DI, page 84]) use a more extended notion of Galois which allows for the presence of idempotents, but here we'll only use 'Galois' in the connected case. Also, since a strongly separable extension is automatically locally strongly separable a Galois extension is an infinite Galois extension. This is unesthetic, but causes no serious confusion.

Definition III.3. A connected commutative ring is *separably closed* if its only connected strongly separable extension is itself. If R is a connected ring then the extension S of R is a *separable closure* of R provided S is also connected, is a locally strongly separable extension of R , and is separably closed.

Definition III.3 applies only to connected rings. The extension to more general rings is complicated and will be done in later chapters.

It is a theorem, due to Harrison, that separable closures always exist. There are several ways to see this: one can start at the bottom with the given base ring and add strongly separable connected extensions until no more can be added, and then the resulting ring is separably closed (the key here is that the process does stop, which is a cardinality argument) or one can start at the top with all the extensions, paste them together freely (i.e., by tensor product) and then discard idempotents until a connected extension, which will be a separable closure, is obtained. The reader will find a careful treatment of the first approach in [DI, pages 99-105]. We will here use the second approach; our reason for doing so is that our subsequent construction of the separable closure in the general case follows the same pattern, and seeing the construction here in the simpler case makes the later construction more natural.

The first step is the following lemma, which also has other applications.

Lemma III.4. Let R be a connected ring and T and S locally strongly separable extensions of R with S also connected. Let $h : T \to S$ be an R-algebra homomorphism. Then the kernel I of h is of the form $I(x)$ for a suitable x in $X(T)$. (See Definition II.14 for the notation.)

Proof: To prove the Lemma, we need to recall the following [DI, Cor. 2.6, page 96]: If $A \to B$ is a homomorphism of separable C-algebras with B a finite projective extension of C then its kernel is generated by an idempotent. Let t be in I and choose strongly separable subextensions T' and S' of T and S such that t is in T' and $h(T')$ is contained in S', and let $I' = I \cap T'$. The above shows that I' is idempotent generated. Thus $t = te$ for some idempotent e in I, and hence I is generated by its idempotents. Since S is connected the set of idempotents in I is a maximal Boolean ideal, and the result follows from II.6.

We can state the result of III.4 as: there is an x in $X(T)$ such that the induced map $T_x \to S$ is one-one.

Proposition III.5. Let R be a connected ring, T a locally strongly separable extension of R and I an ideal of R generated by idempotents. Then T/I is also a locally strongly separable extension of R.

Proof: First consider the case where T is strongly separable over R. T, being a finitely generated projective R-module, can't be written as an infinite direct sum of non-zero submodules. Thus T can be written as $T_1 \times \cdots \times T_n$ where the R-algebras T_i are connected, and T has only a finite number of idempotents. So in this case, if the idempotents $e_1, \ldots, e_k$ generate I then $I = Te$ where $e = e_1 \cup e_2 \cup \cdots \cup e_k$. Thus $T/I = T(1 - e)$ which is a strongly separable R-algebra. In general, write $T = \text{dir lim } T_i$ where each T_i is a strongly separable subextension of T. Let $I_i = I \cap T_i$ and let J_i be the ideal of T_i generated by the idempotents in I_i.

<u>Claim</u>: $I_i = J_i$. For let a be in I_i . Then by II.15 there is an idempotent e of I such that $a = ae$. Choose k such that $T_i \subseteq T_k$ and e is in T_k . Then $e \in J_k$. T_k/J_k is strongly separable over R by the first part of the proof, so by the result of DI quoted in the proof of III.4 above, the kernel of $T_i \to T_k/J_k$ is generated by idempotents. This kernel is contained in I_i and contains a , so a is a linear combination of idempotents in I_i , i.e., $a \in J_i$. This works for any a and hence $I_i \subseteq J_i$. The other inclusion is automatic and the claim follows. Thus $T/I = \text{dir lim } T_i/J_i$ is locally strongly separable.

We next need a result on strongly separable extensions of direct limits, and this will depend on the following characterization of such extensions given by DeMeyer and Ingraham:

(*) The extension A of B is strongly separable if and only if there is a B-linear map $f : A \to B$ and elements $x_1,\ldots,x_n$; $y_1,\ldots,y_n$ of B such that i) $\Sigma x_i y_i = 1$, ii) $\Sigma x_i f(x_i x) = x$ for all x in S [DI, page 92].

<u>Proposition III.6</u>. Let $R = \text{dir lim } R_i$ where each R_i is a subalgebra of R and let S be a strongly separable extension of R . Then (for some ℓ) there is a strongly separable extension S_ℓ of R_ℓ such that $S = R \otimes_{R_\ell} S_\ell$.

Proof: Let $x_1,\ldots,x_n$; $y_1,\ldots,y_n$ and f be as in (*) above. Choose ℓ such that R_ℓ contains all the $f(y_k x_i x_j)$, $f(y_i y_j)$, and $f(x_j)$. Let S_ℓ be the R_ℓ-submodule of S generated by the $x_1,\ldots,x_n$. Since we have $x_i x_j = \Sigma x_k f(y_k x_i x_j)$ and $y_j = \Sigma x_i f(x_i x_j)$ by ii) from (*), S_ℓ is an R_ℓ-algebra containing all the x_i and y_i . Also, f restricted to S_ℓ has image in R_ℓ , so by (*) again (applied to $x_1,\ldots,x_n$; $y_1,\ldots,y_n$ and f restricted to S_ℓ), S_ℓ is a strongly separable extension of R_ℓ . We have a map $R \otimes_{R_\ell} S_\ell \to S$ by multiplication which is onto by ii) of (*). Suppose $\Sigma r_i \otimes s_i$ goes to zero, i.e., $\Sigma r_i s_i = 0$. Then $\Sigma r_i \otimes s_i = \Sigma_i r_i \otimes \Sigma_j x_j f(x_j s_i) = \Sigma_{i,j} r_i \otimes x_j f(x_j s_i) = \Sigma_{i,j} r_i f(x_j s_i) \otimes x_j = \Sigma_j f(x_j \Sigma_i r_i s_i) \otimes x_j = \Sigma_j 0 \otimes x_j = 0$. Thus the map is an isomorphism.

Now we can construct the separable closure. Fix a connected commutative ring R . Let $\{S_i : i \in I\}$ be a set of representatives for the isomorphism classes of strongly separable extensions of R such that each class has countably many representatives in the set. Let $S = \bigotimes_I S_i$. (Recall that the infinite tensor product is defined as follows: let $\mathcal{F} = \{F \subseteq I : F \text{ is finite}\}$. For $F \in \mathcal{F}$, let $S_F = \bigotimes_F S_i$. $\mathcal{F}$ is directed by inclusion and $\bigotimes_I S_i = \text{dir lim } S_F$.)

Note that S is a locally strongly separable extension of R . Pick a point x in $X(S)$. We are going to show that S_x is a separable closure of R . S_x is connected (II.21) and locally strongly separable (III.5). Take a connected strongly separable extension of

S_x . By II.24 the extension is of the form T_x for some strongly separable extension T of S . By III.6 there is an F and a strongly separable extension T_F of S_F such that $S \otimes_{S_F} T_F = T$. Let $S' = \bigotimes_{I-F} S_i$, so $S = S' \otimes_R S_F$ and $T = S' \otimes_R T_F$. We claim that if A is any strongly separable extension of R of rank n , $S' \otimes_R A = S' \times \cdots \times S'$ (n-times). For by [DI, Lemma 2.8, page 97] there is a strongly separable extension S_k of R with $S_k \otimes_R A = S_k \times \cdots \times S_k$. If $S'' = \bigotimes_{i \neq k} S_i$ then $S' = S'' \otimes_R S_k$ and $S' \otimes_R A = S'' \otimes_R S_k \otimes_R A = S'' \otimes_R (S_k \times \cdots \times S_k) = S' \times \cdots \times S'$. Thus, in particular, $S' \otimes_R S_F = S = S' \times \cdots \times S'$ and $S' \otimes_R T_F = T = S' \times \cdots \times S'$. Thus the maps $S' \to S_x$ and $S' \to T_x$ are onto and hence $S_x \to T_x$ is an isomorphism. So S_x has no proper, connected, strongly separable extensions, and is thus separably closed. We have just proven:

<u>Theorem III.7.</u> A connected commutative ring always has a separable closure.

There are a number of properties of separable closures which we will need subsequently and which we now record, with appropriate references to DI: the separable closure is unique up to isomorphism [DI, Theorem 3.3, page 103]; every endomorphism of the separable closure is an automorphism [DI, Corollary 3.4, page 106] and the separable closure of R is an infinite Galois extension of R

[DI, Corollary 3.5, page 106].

We also have the following:

Proposition III.8. Let R be a connected commutative ring, T a locally strongly separable extension of R and S a separable closure of R. Then there is an R-algebra homomorphism $T \to S$.

Proof: First we will show: if A is a connected locally strongly separable extension of S then $A = S$. For A is a direct limit of connected strongly separable extensions of S, and each of these is equal to S itself since S is separably closed. Now let $S' = S \otimes_R T$. S' is a locally strongly separable extension of S. Pick y in $X(S')$. Then, by III.5, S'_y is a (connected) locally strongly separable extension of S. By the above remarks $S'_y = S$. The desired homomorphism $T \to S$ is the composite $T \to S \otimes_R T = S' \to S'_y = S$.

We next turn to the analysis of general locally strongly separable extensions of a connected commutative ring. Such algebras will be described in two parts: their images in a separable closure of the base ring and their idempotents. Proposition III.8 guarantees the existence of the images (we'll need an additional hypothesis to make the image unique) and Lemma III.4 shows that the part of the algebra lost under such images is described by idempotents. The analysis begins with a look at some technical properties of locally strongly separable extensions.

__Definition III.9.__ [DI, page 40]: Let S be a separable algebra over R. The __separability idempotent__ $e = \Sigma a_i \otimes b_i$ of S over R is the (unique) idempotent in $S \otimes_R S$ such that $\Sigma a_i b_i = 1$ and $(s \otimes 1)e = (1 \otimes s)e$ for all s in S. Denote e by $e(T/R)$.

__Proposition III.10.__ Let S be a locally strongly separable extension of R, T a strongly separable subextension of S and e the image in $S \otimes_R S$ of the separability idempotent of T. Then:

(a) S is faithfully T-flat;

(b) $T = \{s \in S : (s \otimes_R 1 - 1 \otimes_R s)e = 0\}$;

(c) The canonical surjection $S \otimes_R S \to S \otimes_T S$ is split by multiplication by e.

__Proof__: Remember that B is a faithfully flat A-algebra if B is a flat A-module such that for any A-module M if $B \otimes_A M = 0$ then $M = 0$. If B is a flat extension of A and B/A is A-flat then B is faithfully A-flat. If B is a faithfully flat extension of A then $A = \{x \in B : x \otimes 1 = 1 \otimes x \text{ in } B \otimes_A B\}$. Now to prove the Proposition, write $S = \text{dir lim } S_i$ where S_i is a strongly separable subextension of S containing T. S_i is then a T algebra and since both S_i and T are strongly separable extensions of R, S_i is a strongly separable extension of T, using [DI, Theorem 2.4, page 94]. So T is a T-direct summand of S_i [DI, Corollary 2.3, page 94] and S_i/T is a projective T-module. Thus $S = \text{dir lim } S_i$ and $S/T = \text{dir lim } S_i/T$ are flat T-modules, being direct limits of flat T-modules and so S is faithfully T-flat. This proves (a). Now map

S to $S \otimes_T S$ by sending s to $s \otimes_T 1 - 1 \otimes_T s$. By part (a), the kernel is T . Now take the (split) exact sequence

$$0 \to J \to T \otimes_R T \overset{\leftarrow}{\to} T \to 0$$

(the map $T \otimes_R T \to T$ sends $a \otimes b$ to ab and the splitting sends t to $t \cdot e(T/R)$) and tensor the sequence over $T \otimes_R T$ with $S \otimes_R S$. We get a split exact sequence

$$0 \to K \to S \otimes_R S \overset{\leftarrow}{\to} S \otimes_T S \to 0$$

where the splitting map sends $x \otimes_T y$ to $(x \otimes_R y) \cdot e$. Thus, for s in S , $s \otimes_T 1 - 1 \otimes_T s = 0$ if and only if $(s \otimes_R 1 - 1 \otimes_R s)\, e = 0$. This and the above remark proves (b), and the second exact sequence is (c).

A separable subextension of a strongly separable extension is itself strongly separable [DI, Proposition 2.5, page 96]. Using Proposition III.10 we can extend this to subextensions of locally strongly separable extensions.

<u>Proposition III.11</u>. Let S be a locally strongly separable extension of R and T a separable subextension. Then T is strongly separable.

<u>Proof</u>: Let $e = \Sigma a_i \otimes b_i$ be the image in $S \otimes_R S$ of $e(T/R)$. Let T' be a strongly separable subextension of S containing all the a_i's and b_i's. Let f be the image in $S \otimes_R S$ of $e(T'/R)$. Then $ef = f$ and if s is in T , $(s \otimes_R 1 - 1 \otimes_R s)f = (s \otimes_R 1 - 1 \otimes_R S)ef =$ so by III.10 (b), s is in T' . Thus $T' \subset T$, and by the result quoted above T is strongly separable.

Next we look at maps of locally strongly separable extensions to infinite Galois extensions. If A and B are any R-algebras we let $\mathrm{Alg}_R(A,B)$ denote the set of all R-algebra homomorphisms from A to B.

__Lemma III.12.__ Let R be connected, T a connected strongly separable extension of R and S an infinite Galois extension of R. Suppose $\mathrm{Alg}_R(T,S)$ is non-empty and give it the discrete topology. Then

$$S \otimes_R T \to C(\mathrm{Alg}_R(T,S),S)$$

(where $(s \otimes t)(h) = sh(t)$) is an S-algebra isomorphism.

__Proof:__ Let G be the group of R-algebra automorphisms of S. First we remark that $\mathrm{Alg}_R(T,S)$ is finite: in fact, by [DI, Corollary 1.6, page 88] if A is any finitely generated separable extension of R and B is a connected extension of R then $\mathrm{Alg}_R(A,B)$ is finite. Let $h : T \to S$ and consider $Y = \{gh : g \in G\}$. By our remark Y is finite, say $Y = \{h_1,\ldots,h_n\}$ where $h_1 = h$. Let $T' = h_1(T) \cdot h_2(T) \cdot \cdots \cdot h_n(T)$. T' is the image in S of $T \otimes_R T \otimes_R \cdots \otimes_R T$ (n-times) and hence is strongly separable. If $g \in G$, $gT' \subset T'$ so we have, by restriction, a homomorphism $G \to \mathrm{Alg}_R(T',T')$; call the image G'. G' is a finite group by our remark above. If a is in $T' - R$, there is a g in G with $ga \neq a$ and if g' is the restriction of g to T', we have $g'a \neq a$. Thus the ring of G' invariants of T' is R itself. This can be expressed by the exactness of the following sequence:

$$(*) \qquad 0 \to R \to T' \to \prod_{i=1}^{n} T'$$

where the last map sends t to $(g_1 t - t, \ldots, g_n t - t)$ where $G' = \{g_1, \ldots, g_n\}$. We let G' act on $T \otimes_R T'$ by $g'(a \otimes b) = a \otimes g'b$. The sequence (*) remains exact after tensoring with the projective module T, and hence

$$0 \to T \to T \otimes_R T' \to \prod_{i=1}^{n} T \otimes_R T'$$

is exact. This means that the ring of G'-invariants of $T \otimes_R T'$ is just T itself. Now let e be a minimal idempotent of $T \otimes_R T'$ and let $e_1 = e, e_2, \ldots, e_k$ be its distinct images under G'. Then $e_1 + \cdots + e_k$ is a G'-invariant and hence an idempotent of T. Since T is connected this idempotent must be one, and so G' acts transitively on the minimal idempotents of T'. There is a map $T' \otimes_R T \to T'$ by $a \otimes b \to ah(b)$. Its kernel is generated by an idempotent f (III.4) and $(T' \otimes_R T)(1 - f)$ is isomorphic to T'. So $1 - f$ is a minimal idempotent. If e is any other minimal idempotent there is a g in G' with $g(1 - f) = e$ and hence $T' = (T' \otimes_R T)(1 - f)$ is isomorphic to $(T' \otimes_R T)e$ via g. Thus we have $T' \otimes_R T = T' \times \cdots \times T'$, the number of factors being equal to the rank of T over R. Thus $S \otimes_R T = S \otimes_{T'} T' \otimes_R T = S \times \cdots \times S$. Let pr_i be the projection on the ith-factor of the product and let $k_i : T \to S$ be $k_i(t) = pr_i(1 \otimes t)$. Since pr_i is an S-algebra homomorphism if $k_i = k_j$ then $pr_i = pr_j$ and so $i = j$. Thus the k_i are all distinct. We know in general [DI, Corollary 1.6, page 88] that there are at most rank-of-A R-algebra homomorphisms from the strongly separable extension A of R to the connected R-algebra B, and so $\{k_i\} = \mathrm{Alg}_R(T,S)$. The map $S \otimes_R T \to \Pi S$ sends $s \otimes t$ to $(sk_1(t),\ldots,sk_n(t))$. But this is precisely the isomorphism we are searching for.

In the course of proving Lemma III.12 we've also obtained some results on homomorphisms which are interesting in themselves and which we now record for future reference.

Corollary III.13. Let R be a connected ring, T a connected strongly separable extension of R and S an infinite Galois extension of R with group of automorphisms G. Then G acts transitively on $\mathrm{Alg}_R(T,S)$.

Proof: In the notation of the proof of III.12, we showed that i) G acts transitively on the minimal idempotents of $T' \otimes_R T$ and ii) the minimal idempotents of $S \otimes_R T$ all lie in $T' \otimes_R T$. This means that G acts transitively on the minimal idempotents of $S \otimes_R T' = C(\mathrm{Alg}_R(T,S),S)$. We continue to let $\mathrm{Alg}_R(T,S) = \{k_1,\ldots,k_n\}$. Let $f = \Sigma a_i \otimes b_i$ be in $S \otimes_R T'$. Then if $h \in \mathrm{Alg}_R(T,S)$ we have $((\sigma \otimes 1)f)(h) = \Sigma\sigma a_i h(b_i) = \sigma(\Sigma a_i \sigma^{-1}h(b_i)) = \sigma(f(\sigma^{-1}h))$. Let e_i be the function $e_i(k_j) = \delta_{ij}$. Each e_i is a minimal idempotent of $C(\mathrm{Alg}_R(T,S),S)$. If $(\sigma \otimes 1)e_i = e_j$, then $\sigma(e_i(\sigma^{-1}k_j)) = 1$ so $e_i(\sigma^{-1}k_j) = 1$ and $\sigma^{-1}k_j = k_i$.

The above result has the following consequence which we will want to use: suppose that $A \subset B$ are strongly separable connected extensions of R and that S is an infinite Galois R-algebra with $\mathrm{Alg}_R(A,S)$ and $\mathrm{Alg}_R(B,S)$ both non-empty. If G is the automorphism group of S over R, G acts transitively on both sets. This means that the natural map $\mathrm{Alg}_R(B,S) \to \mathrm{Alg}_R(A,S)$ by restriction is surjective.

Now suppose $A = \text{dir lim } A_i$ and B are any R-algebras. Then $\mathrm{Alg}_R(A,B) = \text{proj lim } \mathrm{Alg}_R(A_i,B)$. The <u>standard topology</u> on $\mathrm{Alg}_R(A,B)$ will be the projective limit topology induced from the discrete topology on each $\mathrm{Alg}_R(A_i,B)$.

<u>Theorem III.14</u>. Let R be a connected ring, T a locally strongly separable extension of R and S an infinite Galois extension of R. Suppose that for every strongly separable subextension T' of T and every minimal idempotent e of T' that $\mathrm{Alg}_R(eT',S)$ is non-empty. Then the map

$$S \otimes_R T \to C(\mathrm{Alg}_R(T,S),S)$$

(where $(s \otimes t)(h) = sh(t)$) is an S-algebra isomorphism. (Here $\mathrm{Alg}_R(T,S)$ carries the standard topology.)

<u>Proof</u>: First suppose T is a strongly separable extension of R. Let $e_1,\ldots,e_n$ be the minimal idempotents of T and let h_i denote the projection of T on Te_i. The maps h_i induce inclusions $\mathrm{Alg}_R(Te_i,S) \to \mathrm{Alg}_R(T,S)$ and, by III.4, any map $T \to S$ must have some h_i as a factor. Thus $\mathrm{Alg}_R(T,S) = \coprod \mathrm{Alg}_R(Te_i,S)$ (here $\coprod$ is a disjoint union). Then we have isomorphisms

$S \otimes_R T = \Pi(S \otimes_R Te_i) = \Pi C(\mathrm{Alg}_R(Te_i,S),S) = C(\amalg \mathrm{Alg}_R(Te_i,S),S) =$ $C(\mathrm{Alg}_R(T,S),S)$; the second isomorphism coming from III.12. Thus the result holds in case T is strongly separable. In general, $T =$ dir lim T_j where each T_j is strongly separable over R . If $T_j \subset T_k$, then by our remarks above we have $\mathrm{Alg}_R(T_k,S) \to \mathrm{Alg}_R(T_j,S)$ surjective. Thus we have the following chain of isomorphisms:

$C(\mathrm{Alg}_R(T,S),S) = C(\text{proj lim } \mathrm{Alg}_R(T_j,S),S) = \text{dir lim } C(\mathrm{Alg}_R(T_j,S),S) =$ $\text{dir lim } (S \otimes_R T_j) = S \otimes_R T$, the penultimate isomorphism arising from the first part of the proof. This chain of isomorphisms thus gives the theorem.

The special case of the theorem when $T = S$ is also of interest: we let $\mathrm{Aut}_R(S)$ denote the group of R-algebra automorphisms of the R-algebra S .

Lemma III.15. If R is a connected ring and S an infinite Galois R-algebra then $\mathrm{Aut}_R(S) = \mathrm{Alg}_R(S,S)$.

Proof: Let $S =$ dir lim T_i where each T_i is a strongly separable subextension of S . Then $\mathrm{Alg}_R(S,S) = \text{proj lim } \mathrm{Alg}_R(T_i,S)$. By III.13 $\mathrm{Aut}_R(S)$ acts transitively and continuously on each $\mathrm{Alg}_R(T_i,S)$ and hence acts transitively on $\mathrm{Alg}_R(S,S)$. So if h is in $\mathrm{Alg}_R(S,S)$ there is a g in $\mathrm{Aut}_R(S)$ such that $g \cdot id = h$, i.e., $h = g$ is in $\mathrm{Aut}_R(S)$.

Corollary III.16. Let R be a connected ring and S an infinite Galois R-algebra. Then $S \otimes_R S = C(\mathrm{Aut}_R(S), S)$.

Proof: Apply III.14 and III.15.

It is not reasonable to hope that every locally strongly separable extension of the connected ring R is a ring of functions with values in a connected R-algebra (example: R is the real numbers and the algebr is the reals direct sum the complexes). However, there is a condition which does characterize such extensions, and we'll look at that now.

Definition III.17. Let R be a connected ring. An R-algebra T is normal if for every infinite Galois extension S of R all homomorphism from T to S have the same image.

Some simple properties of normal algebras are contained in the following:

Lemma III.18.

i) A homomorphic image of a normal algebra is normal.

ii) A normal, separable subalgebra of an infinite Galois extension is a Galois extension (over the connected base ring R).

Proof: i) Any image of a homomorphic image of the normal algebra is itself an image of the normal algebra, so all such in the same infinite Galois extension are equal.

ii) Let S be the infinite Galois extension, T the normal separable subalgebra and $G = \mathrm{Aut}_R(S)$. Since T is normal, if $g \in G$ then $g(T) = \mathrm{id}(T) = T$. So if $t \in T - R$ and $g(t) \neq t$, the restriction of g to T is in $\mathrm{Aut}_R(T)$ and moves t . T is strongly separable over R by III.11 and the result follows.

We now determine strongly separable normal extensions of a connected ring. First a lemma:

<u>Lemma III.19.</u> Let R be a connected ring, S an infinite Galois R-algebra and T a Galois R-algebra. Then if $h \in \mathrm{Alg}_R(T,S)$, we have $\mathrm{Alg}_R(T,S) = h\,\mathrm{Aut}_R(T)$.

<u>Proof</u>: $T \otimes_R T = C(\mathrm{Aut}_R(T),T)$ by III.16. Then, in an obvious notation, we have $S \otimes_R T = S \otimes_h T \otimes_R T = C(\mathrm{Aut}_R(T), S \otimes_h T) = C(h\,\mathrm{Aut}_R(T),S)$ (the second isomorphism is by I.31). Since also $S \otimes_R T = C(\mathrm{Alg}_R(T,S),S)$ by III.14, the result follows.

<u>Proposition III.20.</u> Let R be a connected ring. A strongly separable extension T of R is normal if and only if T is a finite product of isomorphic Galois extensions of R .

<u>Proof</u>: First suppose T is both normal and strongly separable, and write $T = T_1 \times \cdots \times T_n$ where each T_i is connected. By III.18, each T_i is normal. Let S be a separable closure of R . By III.18, there is homomorphism $h_i : T_i \to S$ which by III.4 is an injection. By III.18 i), $h_i(T_i)$ is normal and by III.18 ii), Galois. Since T is normal,

$h_i(T_i) = h_j(T_j)$ for all i,j and hence $T = h_1(T_1) \times \cdots \times h_1(T_1)$ is a product of isomorphic Galois extensions. Now suppose $T = T_0 \times \cdots \times T_0$ is a finite product of isomorphic Galois extensions and let A be an infinite Galois extension of R. Clearly $\mathrm{Alg}_R(T,A)$ is empty if and only if $\mathrm{Alg}_R(T_0,A)$ is, so we assume neither are and let h be in $\mathrm{Alg}_R(T_0,A)$. By III.19, $\mathrm{Alg}_R(T_0,A) = h\,\mathrm{Aut}_R(T_0)$ and so all images of T_0 in A equal to $h(T_0)$. Since by III.4 every map $T \to A$ factors as $T \to T_0 \to A$, all images of T in A equal $h(T_0)$ also. Thus T is normal.

The reader is warned that a normal subalgebra of a strongly separable algebra need not be invariant under automorphisms of the algebra: let $\mathbb{R}$ be the reals, $\mathbb{C}$ the complexes and let $S = \mathbb{C} \times \mathbb{C} \times \mathbb{C}$. Let $T = \{(a,b,c) \in S : a = b\}$. $T = \mathbb{C} \times \mathbb{C}$ and so is normal over $\mathbb{R}$ by III.20. Let g be the automorphism which cyclically permits the factors of S. Then $(0,0,1) \in T$ but $g(0,0,1) = (1,0,0)$ is not in T.

Of course, normal algebras are the algebras to do Galois theory with and the proper formulation of the notion required some refinement. Normal strongly separable algebras over a connected ring were first studied by Villamayor and Zelinsky (under a different name and characteristic property) in "Galois theory for rings with finitely many idempotents", Nagoya Math. J. 27(1966). These algebras were called weakly Galois algebras; since they're known as such in the literature we'll call them the same thing here - although the name comes from non-commutative simple ring theory. Formally:

Definition III.21. Let R be a connected ring. A strongly separable normal extension of R is a weakly Galois extension of R. An extension of R which is a direct limit of weakly Galois subextensions is a locally weakly Galois extension of R.

Some simple properties of locally weakly Galois algebras that we'll need are in the following lemma:

Lemma III.22. Let R be a connected ring.

i) A locally weakly Galois extension of R is normal.

ii) A connected normal locally strongly separable extension of R is an infinite Galois extension of R.

iii) If T is a locally weakly Galois extension of R so is T_x for x in $X(T)$.

Proof: i) Let $T = \text{dir lim } T_i$, where T_i is a weakly Galois extension of R, be the extension in question and let A be an infinite Galois extension of R. If $h,k \in \text{Alg}_R(T,A)$, $h(T) = \bigcup h(T_i) = \bigcup k(T_i) = k(T)$, the second equality being III.20.

ii) Let T be as in i) and suppose it's connected. Let S be a separable closure of R. By III.8 there is a map $h : T \to S$ which by III.4 is an injection. We can assume, then, that $T = h(T)$. T is normal and so if $g \in \text{Aut}_R(S)$, $g(T) = T$. Then an element of T not in R is moved by some g in $\text{Aut}_R(S)$, and that g restricted to T is in $Aut_R(T)$. So the only $Aut_R(T)$-invariants of T are elements of R.

iii) First suppose T is weakly Galois. Then by III.20, T_x is Galois. Now suppose T is as in i). The image of T_i in T_x is Galois by III.4 and the special case just done, and since T_x is the direct limit of the images of the T_i's, T_x is locally weakly Galois.

Now let T be a locally weakly Galois extension of R and S a separable closure of R. Let T_0 be the image of T in S under an R-algebra homomorphism (by III.22 i) the choice of homomorphism doesn't matter). We'll call T_0 a <u>core</u> of T in S. For every x in $X(T)$ we have (by III.8 and III.5) a map $T_x \to S$ whose image must be T_0. Because T_x is a direct limit of connected strongly separable algebras and T_0 is (by III.20 ii)) an infinite Galois R-algebra, by III.13 and an easy limit argument we have: if h is in $Alg_R(T_x, T_0)$ then $Alg_R(T_x, T_0) = Aut_R(T_0)h$. This leads to:

<u>Proposition III.23.</u> Let R be a connected ring, T a locally weakly Galois extension of R and T_0 a core of T. Then $Aut_R(T_0)$ operates continuously and effectively on $Alg_R(T, T_0)$ and the quotient is homeomorphic to $X(T)$.

<u>Proof</u>: The above discussion covers all points except the homeomorphism assertion. Let $T = \text{dir lim } T_i$, where each T_i is a weakly Galois extension of R. Define $f_i : Alg_R(T_i, T_0) \to X(T_i)$ by $f_i(h) = x$ if $Ker(h) = I(x)$ (see III.4). Let $f = \text{proj lim } f_i$; f sends

$\mathrm{Alg}_R(T,T_0) = \text{proj lim } \mathrm{Alg}_R(T_i,T_0)$ to $X(T) = \text{proj lim } X(T_i)$, and $f(h) = x$ if $\mathrm{Ker}(h) = I(x)$. Since it's the limit of continuous functions, f is continuous. Since for any x in T, $\mathrm{Alg}_R(T_x,T_0)$ is non-empty, f is onto. If $f(h) = f(g) = x$ then there is a k in $\mathrm{Aut}_R(T_0)$ such that $kh = g$ as we noted above. So f induces a continuous bijection between the quotient and $X(T)$ and by compactness this induced map is a homeomorphism.

Now we can prove the main structure theorem for locally weakly Galois algebras over a connected ring, which is an extension to the infinite case of Proposition III.20.

Theorem III.24. Let R be a connected ring. An extension T of R is locally weakly Galois if and only if T is isomorphic to $C(X,S)$ from some profinite space S and some infinite Galois extension S of R.

Proof: Let T_0 be a core of T and $f : \mathrm{Alg}_R(T,T_0) \to X(T)$ the map defined in III.23. By III.21 and I.28, f has a continuous section p. Define a map of T into $C(X(T),T_0)$ by $t(x) = p(x)(t)$. The map is clearly an injection. $C(X(T),T_0)$ is generated as a T-algebra by its idempotents and each of these lie in the image of T, so $T = C(X(T),T_0)$.

For the converse, write $X = \text{proj lim } X_i$, $S = \text{dir lim } S_i$ where the X_i are finite and the S_i Galois extensions of R. Then $C(X,S) = \text{dir lim } C(X_i,S_i)$ and by III.20 each $C(X_i,S_i)$ is weakly

Galois. So $C(X,S)$ is locally weakly Galois.

The description of locally weakly Galois extensions given in III.24 is the first step of the Galois theory of these extensions. The idea is to use III.10, which says that separable subextensions are determined by their separability idempotents, and these idempotents in turn correspond to certain open-closed subsets of the Boolean spectrum. Since we now have essentially determined the relevant Boolean spectra, we can complete the correspondence. We fix the following notations:

Notation III.25. R = connected base ring

S_0 = infinite Galois extension of R

$G = \mathrm{Aut}_R\,(S_0)$

X = profinite space

$S = C(X,S_0)$

By III.24, S is a typical locally weakly Galois extension of R. We first examine $X(S \otimes_R S)$: we have the following chain of isomorphisms (from I.31) $S \otimes_R S = C(X,S_0) \otimes_R C(X,S_0) = C(X,C(X,S_0) \otimes_R S_0) = C(X,C(X,S_0 \otimes_R S_0)) = C(X\mathrm{x}X,S_0 \otimes_R S_0)$. Now by III.16 $S_0 \otimes_R S_0 = C(G,S_0)$ and combining this with the previous isomorphisms gives $S \otimes_R S = C(X\mathrm{x}X\mathrm{x}G,S_0)$. Since we'll want to use it later, we note that the isomorphism is given by $(a \otimes b)(x,y,g) = a(x)g(b(y))$. Now by the results in II, $X(S \otimes_R S) = X\mathrm{x}X\mathrm{x}G$, and, as we remarked above, it is some of the open and closed subsets of this space which fit into the Galois correspondence. This space carries a certain algebraic structure, which we'll now explore.

We recall that a groupoid is a category in which every map is an isomorphism; more precisely,

Definition III.26. A groupoid is a 6-tuple $(M,0,d,r,s,c,i)$ where $M,0$ are sets (of maps and objects, respectively) $d,r : M \to 0$ functions (assigning to a map its domain and range) $s : 0 \to M$ a function (assigning to an object the identity map on that object) $c : M x_{d,r} M \to M$ a function (which composes composable maps) and $i : M \to M$ a function (assigning to a map its inverse). We leave to the reader the task of describing the axioms on the data to guarantee that the maps do in fact behave like their parenthetical descriptions. If M and 0 are finite sets, then the groupoid is a finite groupoid.

We define a morphism of groupoids in the obvious way, and refer to it as both a functor and a (groupoid) homomorphism. It's clear what we mean by a product or an inverse limit of groupoids. We single out for special mention the following case:

Definition III.27. A profinite groupoid is an inverse limit of finite groupoids, where each transition map in the inverse system is surjective.

Note that in a profinite groupoid $(M,0,d,r,s,c,i)$ M and 0 are profinite sets and d,r,s,c,i are continuous maps.

Our space $X \times X \times G$ can be made into a groupoid by taking $X \times X \times G$ for M, X for O, the first projection for d, the second for r, the map $x \to (x,x,1)$ for s, the map $((x,y,g),(y,z,h)) \to (x,z,gh)$ for c and the map $(x,y,g) \to (y,x,g^{-1})$ for i. This is actually a profinite groupoid: write $X = \text{proj lim } X_i$ and $G = \text{proj lim } G_i$ where X_i, G_i are finite and $X \to X_i$ and $G \to G_i$ are onto. Then $X \times X \times G = \text{proj lim } (X_i \times X_i \times G_i)$ (as groupoids). It will turn out that our Galois correspondence involves the subgroupoids of $X \times X \times G$. To be precise, we specify:

<u>Definition III.28</u>. A <u>subgroupoid</u> of a groupoid is a subset, containing all the identities, and closed under composition and inversion. (Since a subgroupoid contains all identities it has the same set of objects as the original groupoid.)

<u>Notation III.29</u>. In Notation III.25, let $G(S/R) = X \times X \times G$. For a subgroupoid H of $G(S/R)$ let $S^H = \{f \in S : g(f(y)) = f(x) \text{ for all } (x,y,g) \text{ in } H\}$. We leave it to the reader to verify that S^H is a subextension of S. For a subextension T of S let $G(S/T) = \{(x,y,g) \text{ in } G(S/R) : g(f(y)) = f(x) \text{ for all } f \text{ in } T\}$. We leave it to the reader to verify that $G(S/T)$ is a subgroupoid of $G(S/R)$.

We can now state the fundamental theorem of Galois theory for locally weakly Galois algebras:

Theorem III.30. Using the Notations III.25 and III.29, the correspondences $H \to S^H$ and $T \to G(S/T)$ induce inverse bijections between the set of all closed subgroupoids of $G(S/R)$ and the set of all locally strongly separable subextensions of S .

The proof is complicated and we'll divide it up into a number of steps.

1. Let T be a strongly separable subextension of S and e the image in $S \otimes_R S$ of $e(T/R)$ (see III.9). Then e is the characteristic function of $H = G(S/T)$, $S^H = T$ and $S \otimes_T S$ is isomorphic to $C(H,S_0)$.

Proof: If $s \in T$, $(s \otimes 1 - 1 \otimes s)e = 0$ so $(s(x) - g(s(y)))e(x,y,g) = 0$ for all $(x,y,g) \in G(S/R)$. Now $e(x,y,g)$ is an idempotent in S_0 , hence either 0 or 1 . By the above equation, if $e(x,y,g) = 1$, $(x,y,g) \in H$. Since, $e = \Sigma a_i \otimes b_i$ $a_i, b_i \in T$, if $(x,y,g) \in H$, $g(b_i(y)) = b_i(x)$ so $e(x,y,g) = \Sigma a_i(x)g(b_i(y)) = \Sigma a_i(x)b_i(x) = (\Sigma a_i b_i)(x) = 1(x) = 1$. So H is the support of e , hence open and closed. By III.10, $T = \{s \in S : (s \otimes 1 - 1 \otimes s)e = 0\} = \{s \in S : (s \otimes 1 - 1 \otimes s) = 0 \text{ on } H\} = S^H$. Finally, since H is open and closed it's profinite, and hence the restriction $S \otimes_R S \to C(H,S_0)$ is onto by I.30 and induces a surjection $S \otimes_T S \to C(H,S_0)$. Suppose f lies in the kernel of this last map. Let g in $S \otimes_R S$ represent f . Then ge also represents f , and ge is the zero function on $G(S/R)$, so is zero. Thus $f = 0$ and the map is an isomorphism.

2. Let H be any open-closed subgroupoid of $G(S/R)$ and let $T = S^H$. Then T is a strongly separable R-algebra and $H = G(S/T)$.

Proof: Let $e = \Sigma a_i \otimes b_i$ be the characteristic function of H. It's an idempotent and, since all identities are in H, $\Sigma a_i b_i = 1$. We have an exact sequence

$$0 \to T \to S \to C(H,S_0)$$

where the second map sends s to $(s \otimes 1 - 1 \otimes s)$ restricted to H. Tensor over R with S (which is R-flat) to get

$$0 \to S \otimes_R T \to S \otimes_R S \to S \otimes_R C(H,S_0) .$$

By I.31 and III.16 $S \otimes_R C(H,S_0) = C(H,C(X,S_0 \otimes_R S_0)) = C(HxX,C(G,S_0)) = C(HxXxG,S_0)$, so our exact sequence becomes

$$0 \to S \otimes_R T \to S \otimes_R S \to C(HxXxG,S_0) ,$$

where the second map becomes: $(s \otimes u)(x,y,g,z,h) = (s \otimes u)(z,x,h) - (s \otimes u)(z,y,hg)$. (It is an instructive exercise to verify this last assertion.) If (x,y,g) is in H, (z,y,hg) belongs to H if and only if (z,x,h) is in H. So $e(x,y,g,z,h) = 0$ for all (x,y,g,z,h) in $HxXxG$, and so e belongs to $S \otimes_R T$. So we can assume that all the b_i are in T. Now let T_1 be a strongly separable subalgebra of S containing all the a_i and b_i, and let e_1 be the image in $S \otimes_R S$ of $e(T_1/R)$. Then, since $ee_1 = e_1$, $T = \{s \in S : (s \otimes 1 - 1 \otimes s)e = 0\} \subseteq T_1 = \{s \in S : (s \otimes 1 - 1 \otimes s)e_1 = 0\}$. The mulitplication map $T_1 \otimes_R T \to T_1$ by $a \otimes b \to ab$ has $a \to (1 \otimes a)e$ as a right T-module inverse. So T_1 is a direct summand of the finitely generated projective T-module $T_1 \otimes_R T$, so is itself T-projective. By [DI, Theorem 2.4, page 94] if a strongly separable algebra is projective as a module over a subalgebra, the subalgebra is strongly separable. This is the situation here and so T is strongly separable.

We now repeat the argument at the end of 1. to see that $S \otimes_T S = C(H,S_0)$. Since also by 1. we have $S \otimes_T S = C(G(S/T),S_0)$, the map $C(H,S_0) \to C(G(S/T),S_0)$ given by restriction is an isomorphism. This implies $H = G(S/T)$.

3. Let T be a locally strongly separable subalgebra of S . Then $H = G(S/T)$ is closed, $T = S^H$, and $S \otimes_T S = C(H,S_0)$.

Proof: $T = \text{dir lim } T_i$ with each T_i strongly separable over R . Each $G(S/T_i)$ is closed by 1., and hence $H = \cap G(S/T_i)$ is closed. Let s be in $S - T$. Let $M_i = \{(x,y,g) \in G(S/T_i) : s(x) \neq g(s(y))\}$. Then the M_i are closed $(M_i = G(S/T_i) - (s \otimes 1 - 1 \otimes s)^{-1}(0)\ )$ and by 1. have the finite intersection property. Thus $M = \cap M_i$ is non-empty. If $(x,y,g) \in M$ then $(x,y,g) \in H$ but $s(x) \neq g(s(y))$. Thus $S^H \subseteq T$; since clearly $T \subseteq S^H$ they are equal. For the final assertion, the isomorphisms $S \otimes_{T_i} S = C(G(S/T_i),S_0)$ of 1. give rise to the isomorphism $S \otimes_T S = \text{dir lim } S \otimes_{T_i} S = \text{dir lim } C(G(S/T_i),S_0) = C(\text{proj lim } G(S/T_i),S_0) = C(G(S/T),S)$.

4. Let H be a closed subgroupoid of $G(S/R)$. Then S^H is locally strongly separable and $H = G(S/S^H)$.

Proof: Using III.31 below, write $H = \cap G_i$, where each G_i is an open-closed subgroupoid of $G(S/R)$. By 2., $G_i = G(S/T_i)$ where

T_i is the (strongly separable) subalgebra S^{G_i}. Let $T = \text{dir lim } T_i$ Then $H = G(S/T)$ and hence, by 3., $S^H = T$ is locally strongly separable.

The proof of Theorem III.30 is now complete, except for the assertion in 4. that a closed subgroupoid of $G(S/R)$ is an intersection of open-closed subgroupoids. The remaining part of this Chapter is devoted to proving this fact, which as it turns out, is somewhat subtle

Suppose we want to prove that a closed subgroup H of the profinite group G is an intersection of open-closed subgroups. Then we write $G = \text{proj lim } G_i$, with G_i finite, and let $p_i : G \to G_i$ be the projection. Then $H = \cap p_i^{-1}(p_i H)$ (since H is closed) and $p_i^{-1}(p_i H)$ is an open-closed subgroup of G containing H. One would like to do the same thing for groupoids, but there is a complication: the inverse image (under a homomorphism) of a subgroupoid is a subgroupoid, but the same is not true for direct images, unless the homomorphism is one-one on objects: (We leave the positive assertions to the reader to verify.) For consider the groupoid which is the disjoint union of a subgroup of order 2 and the one of order 3 of the symmetric group on 3 letters, and map it to the symmetric group by inclusion. This is a homomorphism whose image is not a subgroupoid of its range. Thus we must proceed more carefully.

We want to show that the closed subgroupoid H of the groupoid $X \times X \times G$ is an intersection of open-closed subgroupoids. Let f be in $X \times X \times G - H$. Choose a finite group G' and a continuous homomorphism $G \to G'$ so that the image of f is not in the image of H under the induced map $X \times X \times G \to X \times X \times G'$. This map is one-one on identities and so the image of H is a subgroupoid.

If we can find an open-closed subgroupoid of $X \times X \times G'$ containing the image of H but not that of f, then its inverse image will be an open-closed subgroupoid of $X \times X \times G$ containing H but not f. So it'll suffice to show that the image of H is an intersection of open-closed subgroupoids of $X \times X \times G'$; i.e., we may assume $G = G'$ is finite.

If U and V are subsets of X let $H(U,V) = \{g \in G : (a,b,g) \in H$ for some $a \in U$ and $b \in V\}$. Note that if $U' \subset U$ and $V' \subset V$ then we have $H(U',V') \subset H(U,V)$.

<u>Claim 1</u>. Since H is closed, for all a,b in X, $H(a,b) = \bigcap H(U,V)$ where U,V are neighborhoods of a,b respectively.

<u>Proof</u>: If $a \in U$ and $b \in V$, $H(a,b) \subset H(U,V)$. If g is not in $H(a,b)$, then (a,b,g) is not in H, so a neighborhood of it misses H. Thus there are U,V neighborhoods of a,b with $U \times V \times g$ missing H, i.e., g is not in $H(U,V)$. So the claim follows.

Using the finiteness of G and Claim 1, we can find for each a,b in X neighborhoods $U(a),V(b)$ of a and b such that $H(a,b) = H(U(a),V(b))$. Let $W(a) = U(a) \cap V(a)$. Then $H(a,b) = H(W(a),W(b))$. Refine the open cover $\{W(a)\}$ of X into a partition $\{U_1,\ldots,U_n\}$ of X such that, for each i there is an a_i in U_i with $U_i \subset W(a_i)$

(we can do this since X is profinite). Then $H(a_i,a_j) = H(W(a_i),W(a_j))$. If we are given a,b in X in advance we can fix things so that a,b are among the a_i's. Summing up, we have proven:

<u>Claim 2</u>. Let a,b in X be given. Then there is a partition $\{U_1,\ldots,U_n\}$ of X and elements a_i of U_i such that $H(a_i,a_j) = H(U_i,U_j)$ while $a = a_i$ and $b = a_j$ for some i,j.

Call a partition of the type in Claim 2 <u>H-homogeneous</u>. Let $P = \{U_1,\ldots,U_n\}$ be an H-homogeneous partition with corresponding elements $a_1,\ldots,a_n$. Let $L_{ij} = H(U_i,U_j)$. Since $H(a_i,a_j) \cdot H(a_j,a_k) \subseteq H(a_i,a_k)$, $L_{ij} \cdot L_{jk} \subseteq L_{ik}$. So $\langle P \rangle = \bigcup_{(i,j)} (U_i x U_j x L_{ij})$ is a subgroupoid of $XxXxG$.

<u>Claim 3</u>. Let $P = \{U_1,\ldots,U_n\}$ be an H-homogeneous partition of X with corresponding elements $a_1,\ldots,a_n$. Then $\langle P \rangle$ is an open-closed subgroupoid of $XxXxG$ containing H.

<u>Proof</u>: $\langle P \rangle$ is a subgroupoid and is open-closed. If $(c,d,f) \in H$, with $c \in U_i$ and $d \in U_j$, then since $H(c,d) \subset L_{ij}$ we have (c,d,f) in $U_i x U_j x L_{ij} \subset \langle P \rangle$. So H is contained in $\langle P \rangle$.

<u>Proposition III.31</u>. A closed subgroupoid H of the profinite groupoid $XxXxG$ is an intersection of open-closed subgroupoids.

<u>Proof</u>: As we remarked above, we can assume G is finite. Suppose (a,b,g) is not in H. Choose an H-homogeneous P as in Claim 2 with

$a = a_i$ and $b = a_j$ for some i and j . Then g is not in $H(a_i, a_j) = L_{ij}$, so (a,b,g) is not in $\langle P \rangle$. But by Claim 3, $\langle P \rangle$ is an open-closed subgroupoid containing H , and the result follows.

Bibliographic note on Chapter III.

The material in the Chapter depends strongly on DeMeyer and Ingraham, Separable Algebras over Commutative Rings, Lecture Notes in Mathematics 181, Springer-Verlag, New York, 1971, which also contains an extensive bibliography. The construction of the separable closure "from the top down" appears here for the first time. The structure theorems for locally weakly Galois extensions come from "Galois groupoids", Journal of Algebra 18(1971) and the proof of the fundamental theorem given here is modeled on the proof of a more general fundamental theorem given in that paper - this one is a little more transparent since the full Boolean spectrum machinery doesn't have to be employed. Finally III.31 appears with a false proof in "The separable closure of some commutative rings", Transactions of the American Mathematical Society 170(1972), and the proof here is adopted from the correction to that paper. P. J. Higgins' book, Categories and Groupoids, Von Nostrand-Reinhold, London, 1971, is a good source of algebraic information on groupoids.

IV. The Fundamental Groupoid.

In this Chapter we will describe the category of locally strongly separable extensions of a ring as isomorphic to a certain subcategory of the category of profinite sets on which a profinite groupoid (called the fundamental groupoid of the base ring) acts continuously. As we'll see in the next chapter, sets with groupoid action are occasionally easier to examine than the algebras that give rise to them - this applies in particular to the associated sets arising from subalgebras, which is what Galois theory deals with.

The description is accomplished in two steps: first we examine the case where the base ring is separably closed (and this requires an appropriate definition of separably closed); it turns out that in this case the groupoid is trivial and we're just looking at categories of profinite spaces, where the category isomorphism is easy. Then next we try to reduce the general case to the special one just mentioned by passing from the base ring to its (appropriately defined) separable closure; here we already have the category isomorphism and it's in the attempt to descend that isomorphism back to the original base that the groupoid comes in.

Our first task, then, is the definition and construction of the separable closure. Fix a commutative ring R. Using the construction of the separable closure in the case R is connected (III.7) as a model, what we wish to do is construct an R-algebra S which receives homomorphisms from all strongly separable R-algebras, and so that S

is minimal (in some sense) with respect to this property. Still imitating the connected case, we could start by taking the tensor product S_0 of all strongly separable extensions of R. Then whatever S is we have a map $S_0 \to S$ whose kernel I will be generated by idempotents. So we want an ideal I of S_0 generated by idempotents such that S_0/I is an extension of R (so $I \cap R = 0$) and I is maximal with respect to this property.

If R is connected, this means that $I = I(x)$ for some x in $X(R)$, and we have the separable closure defined in Chapter III. But if R has infinitely many idempotents, some new phenomena may appear. First, it may happen that a locally strongly separable extension of R modulo an ideal generated by idempotents may not be again locally strongly separable, unlike the connected case (Proposition III.5). Next, it may happen that S_0/I has idempotents not in R, again unlike the connected case.

To deal with the first, we first study and characterize extensions of the form T/J, where T is a locally strongly separable extension of R and J an ideal of T generated by idempotents, and require our separable closure to be of this type itself and to receive homomorphisms from all such extensions. Then requiring the separable closure to be minimal with respect to these properties allows us to deal with the second phenomenon and control the new idempotents introduced which appear in the separable closure but not in R: the Boolean spectrum of the separable closure is the Gleason cover (I.22) of the Boolean spectrum of R.

Our study begins, then, with a look at the class of extensions just defined - we describe them first intrinsically, and then recover the form used above.

Definition IV.1. Let R be a ring and S an extension of R . S is a componentially strongly separable extension of R if for each x in X(R) , i.e., for each component x of Spec (R) , S_x is a strongly separable extension of R_x . S is componentially locally strongly separable if for each x in X(R) , S_x is a locally strongly separable R_x-algebra.

It is true (see IV.8 below) but not entirely trivial that "componentially locally strongly separable" equals "locally componentially strongly separable".

The language of Definition IV.1 is a change from the original terms for these objects. "Componentially strongly separable extension" is what was previously known as "separable cover" and "componentially locally strongly separable extension" is what was previously known as "quasi-separable cover". The increase in clarity should justify the increased length.

Strongly separable extensions are obviously componentially strongly separable. To see that the converse is false, we consider an example. Let X be the convergent sequence $\{1,\frac{1}{2},\ldots,0\}$ and $\mathbb{R}$ and $\mathbb{C}$ the reals and complexes with the discrete topology. Let $S = C(X,\mathbb{C})$ and $R = \{f \in S : f(0) \in \mathbb{R}\}$. Then, by the results at the end of

Chapter II, we have $X(R) = X(S) = X$ and for x in X, if $x \neq 0$ $R_x = S_x = \mathbb{C}$ while $R_0 = \mathbb{R}$ and $S_0 = \mathbb{C}$. Thus S is a componentially strongly separable extension of R. Since the rank of S over R is not a continuous function on $X(R)$, S is not a strongly separable extension of R. S is generated, over R by 1 and the constant function $\sqrt{-1}$. So S is not even locally strongly separable over R. Thus the example also shows that a componentially locally strongly separable extension need not be locally strongly separable.

We will study homomorphisms of componentially locally strongly separable extensions. First a lemma:

<u>Lemma IV.2.</u> Let S be an R-algebra and I an ideal of S such that for each x in $X(R)$, I_x is generated by idempotents of S_x. Then I is generated by idempotents of S.

<u>Proof</u>: Let e_0 be an idempotent in I_x. Choose, by II.20, an idempotent e of S such that $e_x = e_0$. Let a be an element of I such that $a_x = e_0 = e_x$. Then by II.16 there is an idempotent e' of R with $e'_x = 1$ such that $ee' = a$. Replace e by $e'e$. Then $e \in I$ and $e_x = e_0$. Let I' be the ideal of S generated by the idempotents of S contained in I. Then we have just shown that $I'_x = I_x$ for all x in $X(R)$, so by II.17 $I' = I$ and the result follows.

<u>Proposition IV.3.</u> Let S and T be componentially locally strongly separable extensions of R and $f : S \to T$ a homomorphism. Then the

kernel of f is idempotent generated.

Proof: Let I be the kernel of f. Then for each x in $X(R)$, I_x is the kernel of f_x, and by IV.2 it's enough to show that each I_x is idempotent generated. But since S_x and T_x are locally strongly separable extensions of R_x, this follows from III.4.

An elementary but closely related result which we now record is:

Proposition IV.4. Let S be a componentially locally strongly separable extension of R and I an ideal of S generated by idempotents such that $I \cap R = 0$. Then S/I is a componentially locally strongly separable extension of R.

Proof: Let x be in $X(R)$. Then I_x is generated by idempotents and S_x is locally strongly separable so by III.5, S_x/I_x is a locally strongly separable extension of R_x. Thus $(S/I)_x = S_x/I_x$ is locally strongly separable, and the result follows.

These propositions combine to give an extrinsic description of componentially locally strongly separable extensions:

Theorem IV.5. Let S be a componentially locally strongly separable extension of R. Then there is a locally strongly separable extension T of S and an ideal I of T generated by idempotents such that $I \cap R = 0$ and T/I is isomorphic to S. Conversely, any such quotient yields a componentially locally strongly separable extension of R.

Proof: For each x in $X(R)$ and each a in S_x we can choose, since S_x is locally strongly separable, a strongly separable subextension T_0 of S_x which contains a. By II.24 there is a strongly separable extension T' of R such that $T'_x = T_0$. By II.25 there is an idempotent e of R with $e_x = 1$ and an Re-algebra homomorphism $h : T'e \to Se$ such that h_x is the inclusion $T_0 \to S_x$. Let $T(a,x) = R(1 - e) \times T'e$. Then we have $f : T(a,x) \to S$ an R-algebra homomorphism such that a is in the image of f_x, while $T(a,x)$ is a strongly separable extension of R. Let $T = \otimes T(a,x)$, the infinite tensor product ranging over all pairs (a,x) as above. Then the tensor product of the f's gives a map $g : T \to S$. Since for each x in $X(R)$ we have by construction that g_x is surjective, g itself is onto. T is locally strongly separable, and by IV.3 the kernel I of g is idempotent generated. This proves half of the Theorem; the converse results immediately from IV.4 and completes the proof.

Definition IV.6. A separable componentially strong extension S of R is a separable extension S of R such that for each x in $X(R)$, S_x is a strongly separable extension of R_x.

"Separable componentially strong" is stricter than "componentially strongly separable": let $X = \{1,\frac{1}{2},\ldots,0\}$ and $\mathbb{R},\mathbb{C}$ the discrete reals and complexes as in the example just before IV.2. If $T = C(X,\mathbb{R})$ and $R = \{f \in C(X,\mathbb{C}) : f(0) \in \mathbb{R}\}$ then $T_x = \mathbb{R}$ for all $x \in X(T) = X$

while $R_x = \mathbb{C}$ if $x \neq 0$ and $R_0 = \mathbb{R}$. So R is a componentially strongly separable extension of T , but it's not separable. If it were, it would be a direct summand of the separable T-algebra $C(X,C) = T \otimes_{\mathbb{R}} \mathbb{C}$ and hence T-projective, but this is not possible since the T-rank function of R is discontinuous at 0 .

We have, however, the following positive results:

Lemma IV.7. A separable subalgebra of a componentially locally strongly separable extension is componentially strong.

Proof: The lemma follows directly from III.11.

Lemma IV.8. A componentially locally strongly separable extension is a direct limit of separable componentially strong subextensions.

Proof: By IV.5, the extension is a homomorphic image of a direct limit of separable algebras, so the extension itself is also a direct limit of separable subalgebras, and now IV.7 completes the proof.

Lemma IV.9. A separable componentially strong extension S of R is a finitely generated R-module.

Proof: DeMeyer and Ingraham [DI, 2.1, page 92] prove that if B is a strongly separable A-algebra with separability idempotent $\Sigma\, c_i \otimes d_i$, then the c_i's generate B as an A-module. So if $e(S/R) = \Sigma\, a_i \otimes b_i$, then for each x in $X(R)$ the $(a_i)_x$ generate S_x as an R_x-module, since S_x is a strongly separable R_x-algebra with separability idempotent $\Sigma\, (a_i)_x \otimes (b_i)_x$. So since the a_i's generate

S at each point of $X(R)$, they generate S globally.

We can now make a useful little refinement in IV.5:

Proposition IV.10. An extension S of R is separable componentially strong if and only if $S = T/I$ where T is a strongly separable R-algebra and I an ideal of T generated by idempotents with $I \cap R = 0$.

Proof: Let T,I be as in the statement of the Proposition. Then $T/I = S$ is separable and, by IV.5 componentially locally strongly separable. So by IV.7 S is componentially strong. Conversely, if S is a separable componentially strong extension then by IV.5 there is a locally strongly separable R-algebra T mapping onto S . By IV.9 S is finitely generated and so we may assume T is also, i.e., T is strongly separable over R . By IV.3 the kernel I of the mapping is idempotent generated, so $S = T/I$ is of the desired form.

The next proposition plays an important role in the construction of the separable closure.

Proposition IV.11. Let T be a componentially locally strongly separable extension of R and S a componentially locally strongly separable extension of T . Then S is a componentially locally strongly separable extension of R .

Proof: Let x be in $X(R)$. We want to show that S_x is a locally strongly separable R_x-algebra. So we assume $R = R_x$ is connected, and T is locally strongly separable over R. By IV.8, $S = \text{dir lim } S_i$ where each S_i is a separable componentially strong extension of T. If each S_i is locally strongly separable over R, so is S, and hence we may assume $S = S_i$ is a separable componentially strong extension of T. By IV.10 there is a strongly separable T-algebra S' and an ideal I of S' generated by idempotents with $I \cap T = 0$ and $S'/I = S$. Write $T = \text{dir lim } T_i$ where each T_i is a strongly separable R-algebra. By III.6 there is an i and a strongly separable extension U_i of T_i such that $S' = U_i \otimes_{T_i} T$. By III.10, T is is T_i-flat and so U_i injects into S'. For each $j \geq i$, let $U_j = U_i \otimes_{T_i} T_j$. Then S' is the direct limit of its subalgebras U_j. Since U_j is strongly separable over T_j and T_j strongly separable over R, we have U_j strongly separable over R by the transitivity of strongly separable extensions [DI, 1.12, page 46]. So S' is strongly separable over R. Since the ideal I of S' is generated by idempotents, this means that, by III.5, $S = S'/I$ is strongly separable over R.

We'll refer to IV.11 as "transitivity of componentially locally strongly separable extensions".

As mentioned above, componentially locally strongly separable extensions arise in considerations of the separable closure. We will now construct the separable closure and use it to classify componentially locally strongly separable extensions. Readers interested in more information about componentially locally strongly separable extensions can find it in "The separable closure of some commutative rings" (see Bibliographic Note).

The separable closure is to receive homomorphic images of all componentially locally strongly separable extensions, and we will begin the investigation by showing that there are such extensions with arbitrary Boolean spectra.

<u>Proposition IV.12</u>. Let R be a ring and $p : Y \to X(R)$ a continuous surjection with Y profinite. Then there is a componentially locally strongly separable extension T of R and a homeomorphism $X(T) \to Y$ such that the diagram

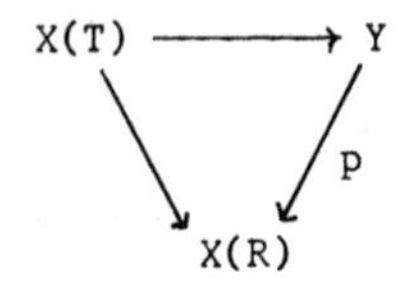

commutes.

<u>Proof</u>: Since we will need the construction of T later it will be done in a functorial manner, and we number the steps in the proof that will be referred to. First, let $T_0(Y) = \prod_y R_{p(y)}$ (the product over all y in Y). Map R to $T_0(Y)$ by $r \to \{r_{p(y)}\}$; this is injective since p is surjective. Let $E(Y)$ be the set of idempotents in $T_0(Y)$ of the following form: to each open-closed set U of Y associate the idempotent $\{e_y\}_{y\varepsilon Y}$ where $e_y = 1$ if $y \varepsilon U$ and zero otherwise. Let $T(Y)$ be the R-subalgebra of $T_0(Y)$ generated

by $E(Y)$. If $y \varepsilon Y$ let $e(y)$ be the element of $T_0(Y)$ whose y^{th}-component is 1 and whose other components are zero. If M denotes the set of idempotents of $T_0(Y)$ whose product with $e(y)$ is zero, then M is easily seen to be a maximal Boolean ideal and hence corresponds to a point of $X(T_0(Y))$. So we have an injection $Y_d \to X(T_0(Y))$ (where Y_d is Y with the discrete topology) and we identify Y_d with its image in $X(T_0(Y))$. Under this identification the map $T_0(Y) \to T_0(Y)_y$ is just projection on the yth-factor of the product. Let $q : X(T_0(Y)) \to X(R)$ be the induced map.

Claim 1. If e is an idempotent of $T_0(Y)$ and x is in $X(R)$ with $e_x \neq 0$, there is a y in $p^{-1}(x)$ with $e_y = 1$.
Proof: Let $F = \{z \varepsilon q^{-1}(x) : e_z = 0\}$. Then $F = q^{-1}(x) \cap N(1 - e)$ is closed. But the set $p^{-1}(x)$ is dense in $q^{-1}(x)$ (this is because $X(T_0(Y))$ is the Stone-Cech compactification of the injection $Y_d \to X(T_0(Y))$). So, since F is closed, if $p^{-1}(x) \subseteq F$, $q^{-1}(x) = F$. But this means that $e_z = 0$ for all z in $q^{-1}(x)$, so $e_x = 0$ also.

Claim 2. Let e be an idempotent of $T_0(Y)$ and x in $X(R)$. If e_x is not zero, the R_x-annihilator of e_x consists of zero alone.
Proof: Since $e_x \neq 0$, there is by Claim 1 a y in $p^{-1}(x)$ with $e_y = 1$. Then $R_x \to T_0(R)_y = R_x$ becomes the identity, so if $r_x e_x = 0$, $r_x e_y = r_x = 0$.

Claim 3. Let E be a set of idempotents of $T_0(Y)$ closed under products and unions and containing all the idempotents of R. Then every idempotent in RE belongs to E.

Proof: Suppose $f = \Sigma r_i e_i$, $r_i \in R$, $e_i \in E$ is an idempotent of E. We can assume that $e_i e_j = 0$ if $i \neq j$ (the general formula is complex so we indicate this simplification for two terms only: $r_1 e_2 + r_2 e_2 = r_1(e_1 - e_1 e_2) + (r_1 + r_2)e_1 e_2 + r_2(e_2 - e_1 e_2)$). Then $f^2 = f$ implies that $(r_i e_i)^2 = r_i e_i$ for each i. So we may assume there is only one summand and $f = re$. First, we show that for each x in $X(R)$ there is an idempotent $g(x)$ of E with $g(x)_x = f_x$. If $e_x = 0$, $g(x) = 0$ works. If $e_x \neq 0$, $r_x^2 e_x = r_x e_x$ implies, by Claim 2, that $r_x^2 = r_x$. Using II.20, choose an idempotent s of R with $s_x = r_x$. Then let $g(x) = se$. Since $g(x)_x = f_x$, there is an idempotent e with $x \in N(e)$ such that $g(x)_y = f_y$ for all y in $N(e)$, by II.16. These neighborhoods cover $X(R)$ and we refine them to a partition $N(e_1), \ldots, N(e_n)$. Choose for each i x_i in $N(e_i)$ so that $g(x_i) = f$ on $N(e_i)$. Then $g = \Sigma e_i g(x_i)$ is an element of E with $g_x = f_x$ for all x in $X(R)$, so $g = f$.

Claim 4. $X(T(Y))$ can be canonically identified with Y.

Proof: By Claim 3, the set of idempotents of $T(Y)$ is just $E(Y)$, and now $E(Y)$ (as a Boolean ring) determines $X(T(Y))$. But $E(Y)$ is just the Boolean algebra of open-closed subsets of the profinite space Y, and hence its maximal ideal space is, by the Stone Representation Theorem, just Y again.

Claim 5. Let E be any set of idempotents of $T_0(Y)$. Then $S = RE$ is a componentially locally strongly separable extension of R .

Proof: Let x be in $X(R)$; we want to show that S_x is a locally strongly separable extension of R_x . Suppose e_0 is a non-zero idempotent of S_x . By II.20 there is an idempotent e of S with $e_x = e_0$ and so by Claim 2 e_0 is R_x-torsion free. Let F be the set of all idempotents of S_x (so $S_x = R_xF$) and write $F = \text{dir lim } F_i$ where each F_i is finite, contains zero and one, and is closed under products and complements. Then $S_x = \text{dir lim } R_xF_i$, so it will suffice to show that each R_xF_i is strongly separable, i.e., we may assume $F = F_i$ is finite. Let $e_1,\ldots,e_n$ be the minimal idempotents of F . Then $R_xF = \Pi R_xe_i$, and since each e_i is R_x-torsion free, we have $R_xF = \Pi R_x$ is strongly separable.

Claim 5 tells us that $T(Y)$ is componentially locally strongly separable. By Claim 4, $X(T(Y)) = Y$ and so $T(Y)$ is the extension we wished to produce. This completes the proof of the proposition.

In the course of proving Claim 5, we see that we actually established the following lemma, which has an interesting consequence:

Lemma IV.13. Let R be a connected ring and S an R-algebra such that S is generated over R by idempotents and every non-zero idempotent of S is R-torsion free. Then $S = C(X(S),R)$.

Proof: In Claim 5 we proved: $S = \text{dir lim } S_i$, where $S_i = \Pi R$. Thus S_i is weakly Galois (III.21) and S is locally weakly Galois (III.21).

Since S is generated over R by idempotents, the homomorphic image of S in a separable closure of R is just R itself. So the lemma follows from III.24.

<u>Corollary IV.14</u>. Let every element of R be idempotent (i.e., R is a Boolean ring). Then $R = C(X(R), Z/2Z)$.

<u>Proof</u>: R is a $Z/2Z$-algebra and every non-zero element of R is $Z/2Z$-torsion free, so IV.13 applies.

The corollary is, of course, part of the Stone Representation Theorem, which gives a category anti-equivalence between the category of Boolean rings and the category of profinite spaces. Using IV.14 and the results on function rings at the end of Chapter I, the reader can construct a complete proof of the theorem.

Proposition IV.12 puts some restrictions on the Boolean spectrum of rings which receive homomorphisms from all their componentially locally strongly separable extensions, as we'll now see. First a definition:

<u>Definition IV.15</u>. A ring S is <u>separably closed</u> if for every componentially locally strongly separable extension T of S there is an S-algebra homomorphism $T \to S$.

<u>Theorem IV.16</u>. A ring S is separably closed if and only if the following conditions are satisfied:

i) $X(S)$ is extremely disconnected (see Definition I.15).

ii) For each x in $X(S)$, S_x is separably closed in the sense of Definition III.3.

Proof: First suppose S is separably closed. Let $p : Y \to X(S)$ be a continuous surjection of topological spaces. Choose T as in IV.12. Then there is an S-homomorphism $T \to S$; since $S \to T \to S = id$, we have induced continuous maps $X(S) \to X(T) \to X(S) = id$. By IV.12 the second map can be identified with p, i.e., p has a section. But if every surjection to $X(S)$ with profinite source has a section, $X(S)$ is extremely disconnected, whence i). If T_0 is a connected strongly separable S_x-algebra for x in $X(S)$, by II.24 there is a strongly separable S-algebra T with $T_x = T_0$. Then the S-homomorphism $T \to S$ induces an S_x-homomorphism $T_0 = T_x \to S_x$, so $T_0 = S_x$. Thus S_x has no proper connected strongly separable extensions, and so is separably closed.

Next suppose that S satisfies conditions i) and ii). Let T be a componentially locally strongly separable extension of S. Then the continuous function $X(T) \to X(S)$ has a section whose image is a closed subset F of $X(T)$ homeomorphic to $X(S)$. $F = \cap N_T(e_i)$ for some collection e_i of idempotents of T. Let I be the ideal of T generated by all $1 - e_i$. Then the map $T \to T/I$ induces an injection $X(T/I) \to X(T)$ whose image is F. So $X(T/I) \to X(S)$ is a bijection. This means that for each x in $X(S)$, $(T/I)_x$ is connected. $(T/I)_x$ is also locally strongly separable and since S_x is separable closed, $(T/I)_x = S_x$. This holds at each x so

$T/I = S$. Then the composite $T \to T/I \to S$ is the desired S-algebra homomorphism.

The separable closure of a ring should be the "smallest" separably closed componentially locally strongly separable extension of the ring, where smallest is in the following sense: if S_1 and S_2 are two such extensions there is a map $S_1 \to S_2$. If S_1 is as small as possible, this map can't have a kernel. Thus we make a definition:

Definition IV.17. A componentially locally strongly separable extension S of the ring R is *minimal* if every homomorphism from S to a componentially locally strongly separable extension of R is a monomorphism.

Proposition IV.18. Let S be a componentially locally strongly separable extension of the ring R . S is minimal if and only if the induced map $X(S) \to X(R)$ is a minimal map (see I.16).

Proof: For any componentially locally strongly separable extension T of R and homomorphism $S \to T$, the kernel is generated by idempotents (IV.3) and meets R in zero. Conversely, if I is an ideal of S generated by idempotents and meeting R in zero, S/I is a componentially locally strongly separable extension of R . Ideals I of S generated by idempotents correspond to closed subsets $V(I) = \bigcap \{N(e) : 1 - e \in I\}$ of $X(S)$. Thus to prove the proposition, we will show that $V(I)$ maps onto $X(R)$ if and only if $I \cap R = 0$. We have

$X(S/I) \to X(S) \to X(R) = X(S/I) \to X(R/I \cap R) \to X(R)$ and the image of the first map is $V(I)$. So $V(I)$ maps onto $X(R)$ if and only if $X(R/I \cap R) = X(R)$; i.e., for all x in $X(R)$, $R_x \neq I_x \cap R_x$. Replace R by R_x, S by S_x and I by I_x. If $I \cap R \neq R$, I is contained in a prime ideal P of S, and if y is the component in which P lies, we have by II.3 that $I \subseteq I(y)$. Thus $I \cap R \subseteq I(y) \cap R = (0)$ since R is connected. The proposition now follows.

Definition IV.19. A separable closure of R is a minimal, separably closed, componentially locally strongly separable extension of R.

Theorem IV.20. A ring R always has a separable closure, and any two such are R-isomorphic.

Proof: For existence, we first do the case where $X(R)$ is extremely disconnected. In this case, let $\{S_i : i \in I\}$ be a set of representatives of the isomorphism classes of strongly separable R-algebras, so that each isomorphism class is represented at least countably many times. Let $S' = \bigotimes_I S_i$. The induced map $X(S') \to X(R)$ has a section whose image is a closed subset $V(J)$ corresponding to an ideal J of S' generated by idempotents (as above) and $X(S'/J) \to X(R)$ is a homeomorphism. Let $S = S'/J$. Since S' is locally strongly separable and J is generated by idempotents, by IV.4, S is a componentially locally strongly separable extension of R, and $X(S) = X(R)$ is extremely disconnected. To show S separably closed, therefore, we use IV.16: i.e., we need: if x is in $X(S)$, S_x is separably close Think of x as being in $X(R)$. Then since J_x is generated by idempotents and $S_x = S_x'/J_x$ has no non-trivial idempotents, $J_x = I(y)$ for some y in $X(S_x')$. Thus $S_x = (S_x')_y$. Now S_x' is a tensor product of strongly separable extensions of R_x , and by II.24 every isomorphism class of strongly separable R_x-algebra is represented at least countably many times. So by III.7, $(S_x')_z$ is a separable closure of R_x for each z in $X(S_x')$, in particular, this holds if $z = y$.

Now suppose R is arbitrary, and let $Y \to X(R)$ be the Gleason cover (I.22) of $X(R)$. Let $T = T(Y)$ be the componentially locally strongly separable extension constructed in IV.12 with $X(T) = Y$. Thus $X(T)$ is extremely disconnected. Let S be the closure of T constructed in the first paragraph. By transitivity of componentially locally strongly separable extensions (IV.11), S is such an extension of R . We've already shown S to be separably closed, and the map $X(S) = X(T) = Y \to X(R)$ is a minimal map, so IV.18 shows that S is indeed a separable closure.

Finally, let S and S' be separable closures of R . Then $S \otimes_R S'$ is a componentially locally strongly separable extension of S , so there is an S-homomorphism $S \otimes_R S' \to S$ and hence an R-homomorphism $g : S' \to S$. This induces a continuous function $f : X(S) \to X(S')$ which is a homeomorphism by I.17 since both $X(S)$ and $X(S')$ are Gleason covers of $X(R)$. Let $y \in X(S)$; then if $f(y) = x$, the induced R_x-homomorphism from $S'_{f(y)} \to S_y$ is a homomorphism of separable closures, so by the result of DeMeyer and Ingraham

[DI, 3.4, page 106] quoted above (page III-7) the map is an isomorphism. It follows that g itself is an isomorphism.

We want to study componentially locally strongly separable extensions of a ring R, and we begin with the case where R is separably closed. First we look at the connected case:

<u>Lemma IV.21</u>. Let R be a separably closed ring containing no non-trivial idempotents. Then an extension S of R is locally strongly separable if and only if for each y in $X(S)$, $S_y = R$.

<u>Proof</u>: If S is locally strongly separable, so is S_y, and since S_y is also connected and R is separably closed, $R = S_y$. Suppose conversely that $S_y = R$ for all y in $X(S)$. This implies that S is generated over R by idempotents. If e is an idempotent of S and $re = 0$, either $e = 0$ or there is a $y \in X(S)$ with $e_y = 1$. Then $re_y = r = 0$ in $S_y = R$ in the latter case, so $r = 0$ and we conclude that non-zero idempotents of S are R-torsion free. By IV.13 $S = C(X(S),R)$ which by III.24 is locally strongly separable over R.

Actually IV.21, its proof, and IV.13 give us a little more: the categories $\mathcal{S}$ of R-algebras S such that $S_y = R$ for all y in $X(S)$ and $\mathcal{P}$ of profinite spaces X are anti-equivalent under the functors $S \to X(S)$ and $X \to C(X,R)$ (here R is still separably closed and connected). We wish to extend this equivalence to general R. We begin by extending IV.21.

Proposition IV.22. Let R be a separably closed ring. Then an extension S of R is componentially locally strongly separable if and only if for each y in $X(S)$ the natural map $R \to S_y$ is surjective with idempotent-generated kernel.

Proof: If S is componentially locally strongly separable, and $y \in X(S)$, then choose x in $X(R)$ so that $R_x \to S_y$. Since S_x is locally strongly separable and $S_y = (S_x)_y$, by IV.21 $S_y = R_x$ and $R \to S_y$ is onto with kernel $I(x)$. Conversely, suppose $R \to S_y$ is surjective with idempotent-generated kernel for all $y \in X(S)$. Choose $x \in X(R)$. Then for all $y \in X(S_x)$, $R_x \to (S_x)_y$ is bijective so by IV.21 S_x is locally strongly separable. So S itself is componentially locally strongly separable.

Now we'll prove the general category isomorphism mentioned above.

Notation IV.23. R is any ring. $\mathcal{S}(R)$ = the category whose objects are extensions S of R such that for each y in $X(S)$ the map $R \to S_y$ is onto with idempotent-generated kernel, and whose morphisms are R-algebra homomorphisms. $\mathcal{P}(R)$ = the category whose objects are pairs (X,p) where X is a profinite space and $p : X \to X(R)$ is a continuous surjection. A morphism $g : (X,p) \to (Y,q)$ of such pairs is a continuous map $f : X \to Y$ such that $qf = p$.

Theorem IV.24. Let R be a ring. Then the categories $\mathcal{S}(R)$ and $\mathcal{P}(R)$ are anti-equivalent.

<u>Proof</u>: The proof is a little complicated and so we'll break it up into several steps. First we need to define functors between the categories. If $S \varepsilon \mathcal{S}(R)$, then the structure map $p : R \to S$ defines a pair $(X(S),X(p))$ in $\mathcal{P}(R)$, which we denote $F(S)$, and if $f : S \to S'$ in $\mathcal{S}(R)$ then we have $X(f) : F(S') \to F(S)$ in $\mathcal{P}(R)$, so letting $F(f) = X(f)$ makes F a functor from $\mathcal{S}(R)$ to $\mathcal{P}(R)$. Now suppose (Y,p) is an object of $\mathcal{P}(R)$. Then the algebra $T(Y)$ defined in the proof of IV.12 lies in $\mathcal{S}(R)$. If $g : (Y,p) \to (X,q)$ is a map in $\mathcal{S}(R)$, we define a homomorphism $T(g) : T(X) \to T(Y)$ as follows: first we define $T_0(g) : T_0(X) \to T_0(Y)$ by writing $T_0(X) = \prod_{z \varepsilon X(R)} \prod_{x \varepsilon q^{-1}(z)} R_{q(z)}$ and $T_0(Y) = \prod_{z \varepsilon X(R)} \prod_{x \varepsilon q^{-1}(z)} \prod_{y \varepsilon g^{-1}(x)} R_{p(y)}$ and then sending each $R_{q(z)}$ to $\prod_{y \varepsilon g^{-1}(x)} R_{p(y)}$ diagonally. Then $T_0(g)(E(X)) \subseteq E(Y)$ so $T_0(g)$ induces a map $T(g) : T(X) \to T(Y)$. (The notations $T_0(\cdot)$ and $E(\cdot)$ are defined in the proof of IV.12.) Thus T is a functor from $\mathcal{P}(R)$ to $\mathcal{S}(R)$ - note that both F and T are contravariant.

<u>Claim 1</u>. $T(F(\cdot))$ is naturally equivalent to the identity.

<u>Proof</u>: Let $S \varepsilon \mathcal{S}(R)$. Then $T_0(X(S)) = \prod_{X(S)} S_y$, and we have a monomorphism $S \to T_0(X(S))$ by the diagonal. The image contains $T(X(S))$, and in fact they are equal since S is generated over R by idempotents (this latter since it holds by IV.21 for S_x for all $x \varepsilon X(R)$). So we have an isomorphism $S \to T(F(\cdot))$. It's easy to see this is natural, and hence Claim 1 holds.

<u>Claim 2</u>. $F(T(\cdot))$ is naturally equivalent to the identity.

Proof: Let $(Y,p) \in P(R)$. Then by IV.12 we have a canonical homeomorphism $g : Y \to X(T(Y))$ such that $g : (Y,p) \to F(T(Y))$ is an isomorphism in $P(R)$. Again it's easy to check that this isomorphism is natural, and Claim 2 holds.

Claims 1 and 2 then prove IV.24.

With IV.22 and IV.24 we have now described the category of componentially locally strongly separable extensions of a separably closed ring R . So we can partially describe the category when R is arbitrary: first embed R in its separable closure S , using IV.20. Then $S \otimes_R (\cdot)$ embeds the category in the similarly defined category over S , and this latter is, we now know, anti-equivalent to the category of profinite spaces surjectively over $X(S)$. So to describe the part of the category which comes from extensions of R , we'll show that the profinite sets in this part of the category carry an additional structure (which turns out to be a continuous groupoid action) which characterizes them. We begin constructing this action now. If $f : A \to B$ is a ring homomorphism and $x \in X(B)$, we denote $X(f)(x)$ by $x \cap A$.

Lemma IV.25. Let R be a ring with separable closure S , and let T be a componentially locally strongly separable extension of R . Then $X(S \otimes_R T) = \{(a,b,h) : a \in X(S) , b \in X(T) , a \cap R = b \cap R$ and $h : T_b \to S_a$ is an R-algebra embedding$\}$.

Proof: Let $z \in X(S \otimes_R T)$. Let $a = z \cap S$, $b = z \cap T$ and

$x = z \cap R$. Then $(S \otimes_R T)_z = (S_a \otimes_{R_x} T_b)_z$. Now $S_a = (S_x)_a$ is a locally strongly separable connected extension of R_x which is separably closed since S is. Then since $T_b = (T_x)_b$ is locally strongly separable over R_x , $S_a \otimes_{R_x} T_b$ is locally strongly separable over S_a and hence by IV.21 $(S \otimes_R T)_z = (S_a \otimes_{R_x} T_b)_z = S_a$. Also $T_b \to (S \otimes_R T)_z$ is a monomorphism; we define $h : T_b \to S_a$ by $(h(y) \otimes 1)_z = (1 \otimes y)_z$ for $y \in T_b$ (i.e., $T_b \to (S \otimes_R T)_z = S_a$ is h). So to each z in $X(S \otimes_R T)$ corresponds a triple of the type in the lemma.

Now suppose (a,b,h) is such a triple. Let $x = a \cap R$ $(= b \cap R)$ and consider the map $g : S_a \otimes_{R_x} T_b \to S_a$ by $g(s_a \otimes t_b) = s_a h(t_b)$. Since both the range and domain of g are locally strongly separable R_x-algebras the kernel of g is idempotent-generated and since S_a is connected the kernel is of the form I(w) for some w in $X(S_a \otimes_{R_x} T_b)$. Let z be the image of w under $X(S_a \otimes_{R_x} T_b) \to X(S \otimes_R T)$, so I(z) is the kernel of the surjection $S \otimes_R T \to S_a$ by $s \otimes t \to s_a h(t_b)$. Thus to each triple corresponds a point in $X(S \otimes_R T)$. We want to show that these correspondences are mutually inverse.

If we start with z and produce a triple (a,b,h) as above, the point z' corresponding to (a,b,h) is such that I(z') is the kernel of $g : S \otimes_R T \to S_a$ by $g(s \otimes t) = s_a h(t_b)$. We have the commutative diagram

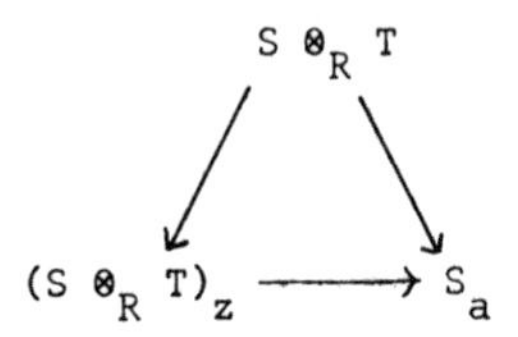

(where the left map is the canonical one, the right one g and the bottom the isomorphism constructed above), so the kernels of both vertical maps are equal, i.e., $I(z) = I(z')$ so $z = z'$.

Now start with (a,b,h) and produce a point z . We look at the associated triple (a',b',h') . $I(a')$ is the kernel of $S \to (S \otimes_R T)_z$ S_a , i.e., $a' = a$. $I(b')$ is the kernel of $T \to (S \otimes_R T)_z$, so $e \in$ if and only if $(1 \otimes e)_z = 0$, which means $1_a h'(e_b) = 0$, or $e_b = 0$, so $I(b') = b$. Finally for t in T , $(1 \otimes t_b)_z = (h(t_b) \otimes 1)_z$; the first is sent to $h'(t_b)$ in S_a and the second to $h(t_b)$ under the isomorphism $(S \otimes_R T)_z \to S_a$, so $h' = h$. This completes the proo of IV.25.

<u>Corollary IV.26</u>. Let R be a ring with separable closure S . Then $X(S \otimes_R S) = \{(a,b,g) : a,b \in X(S) \text{ and } a \cap R = b \cap R \text{ and } g : S_b \to S_a \text{ is an isomorphism}\}$.

We can partially multiply triples (a,b,g) and (b,c,h) as in IV.26 by calling (a,c,gh) their product. This in fact makes the set of triples into a groupoid (III.26) with set of objects $X(S)$, range map the first projection, domain map the second, with the identity at a being $(a,a,1)$ and $(a,b,g)^{-1} = (b,a,g)$. In fact, this groupoid structure arises from the ring structure, as we'll see shortly.

Lemma IV.27. Let R be a ring with separable closure S and let T be a componentially locally strongly separable extension of R. Then the map $h : S \otimes_R T \to (S \otimes_R S) \otimes_S (S \otimes_R T)$ induces a map $X(S \otimes_R S) \times_{X(S)} X(S \otimes_R T) \to X(S \otimes_R T)$ which, in terms of the triples of IV.25, sends $((b,a,g),(a,d,h))$ to (b,d,gh).

Proof: First, the domain of $X(h)$ is actually $X((S \otimes_R S) \otimes_S (S \otimes_R T))$, but by IV.24 (and IV.22) this is homeomorphic with $X(S \otimes_R S) \times_{X(S)} X(S \otimes_R T)$. The rest of the proof consists in unwinding some identifications. Given the triples (b,a,g) and (a,d,h), send $(S \otimes_R S) \otimes_S (S \otimes_R T)$ to S_b by sending $s \otimes u \otimes v \otimes t$ to $s_b g(u_a v_a) gh(t_d)$. As before, the kernel of this is $I(w)$ for some $w \in X((S \otimes_R S) \otimes_S (S \otimes_R T))$. The inclusion on the first factor $S \otimes_R S \to (S \otimes_R S) \otimes_S (S \otimes_R T)$ gives the map $S \otimes_R S \to S_b$ by $s \otimes u \to s_b g(u_a)$, so the projection of w on $X(S \otimes_R S)$ is (b,a,g). The inclusion on the second factor $S \otimes_R T \to (S \otimes_R S) \otimes_S (S \otimes_R T)$ yields the map $S \otimes_R T \to S_b$ by $u \otimes t \to g(u_a) gh(t_d)$. The kernel of this is the same as the kernel of $S \otimes_R T \to S_a$ by $u \otimes t \to u_a h(t_d)$, so the projection of w on $X(S \otimes_R T)$ is (a,d,h). Finally preceding the map by h gives $S \otimes_R T \to S_b$ by $s \otimes t \to s_b gh(t_d)$, i.e., $X(h)(w) = (b,d,gh)$, and the lemma follows.

Corollary IV.28. Let R be a ring with separable closure S. Then $X(S \otimes_R S)$ is a topological groupoid (where the partial multiplication is, in terms of triples, that described above).

Proof: By IV.28 with $T = S$, the triple multiplication is continuous. The range and domain maps result from $X(\cdot)$ applied to the inclusions $S \to S \otimes_R S$ on the first and second factors, respectively, and so are continuous. We leave to the reader the verification that the maps $S \otimes_R S \to S$ by multiplication and $S \otimes_R S \to S \otimes_R S$ by switching factors give, upon applying $X(\cdot)$, the identity section and the inversion map.

The map of IV.27 now makes $X(S \otimes_R T)$ into a set on which the groupoid $X(S \otimes_R S)$ acts, in a sense which we now make precise.

Definition IV.29. Let $(M,0,d,r,s,c,i)$ be a groupoid (III.26). Let A be a set and $p : A \to 0$ a function. Let $q : M \times_{d,p} A \to 0$ be projection on the first factor followed by r. An action of the groupoid on (A,p) is a function $d : M \times_{d,p} A \to A$ satisfying:

i) $pd = q$

ii) if $r(g) = d(f) = p(a)$, $d(f,d(g,a)) = d(fg,a)$

iii) if f is the identity at $p(a)$, $d(f,a) = a$.

If, as usual, we denote the groupoid by its set of morphisms M, then we say that A above is an M-set. If (A,p) and (B,k) M-sets and $h : A \to B$ is a function with $kh = p$, h is an M-set morphism if the diagram

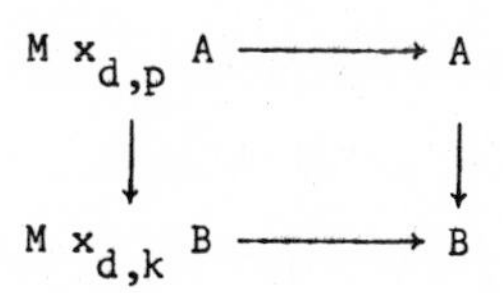

commutes, where the vertical maps are $1 \times h$ and h and the horizontal maps come from the M-action.

Finally if M is a topological groupoid, A a topological space, and d and p continuous functions, then (A,p) is a topological M-set, or space with continuous M-action.

We use the language of IV.29 to interpret IV.27.

Proposition IV.30. Let R be a ring with separable closure S and let T be a componentially locally strongly separable extension of R. Let $p : S \to S \otimes_R T$ be the inclusion on the first factor. Then $(X(S \otimes_R T), X(p))$ is a space with continuous $X(S \otimes_R S)$-action.

Proof: Let the action map d be the map of IV.27, which is continuous. The verification of i), ii), iii) of IV.29 is now easy, using the interpretation of d in terms of triples given in IV.29.

If R is a ring with separable closure S then by IV.30 the functor $X(S \otimes_R (\cdot)) : \mathcal{S}(R) \to \mathcal{P}(S)$ (see Notation IV.23) has range in the subcategory of $\mathcal{P}(S)$ consisting of pairs (A,p) with continuous $X(S \otimes_R S)$-action. In fact we have more:

<u>Theorem IV.31</u>. Let R be a ring with separable closure S. Let $\mathcal{S}(R)$ denote the category of componentially locally strongly separable extensions of R, and let $\mathcal{M}(X(S \otimes_R S))$ denote the category of pairs (A,p) (where A is profinite and $p : A \to X(S)$ is a continuous surjection) with continuous $X(S \otimes_R S)$-action. Then $X(S \otimes_R (\cdot)) : \mathcal{S}(R) \to \mathcal{M}(X(S \otimes_R S))$ is a (contravariant) equivalence of categories.

<u>Proof</u>: We recall that a functor is an equivalence of categories if it fully faithful (i.e., is a bijection on hom-sets) and if every object i the range category is isomorphic to the image of some object in the domain category, and we'll verify these two things here. The second is the more difficult, and we begin with it: suppose (A,p) is an object of $\mathcal{M}(X(S \otimes_R S))$. Let $d : X(S \otimes_R S) \times_{X(S)} A \to A$ be the continuous $X(S \otimes_R S)$-action. Also we have $q : X(S \otimes_R S) \times_{X(S)} A \to X(S)$ (q is the map of Definition IV.29), and IV.29 i) shows that d is a map of sets over $X(S)$. Let $T = T(A)$ be the componentially locally strongly separable extension of IV.12 such that $X(T) = A$. Then by IV.24 there is a unique map $\phi : T \to (S \otimes_R S) \otimes_S T$ such that $X(\phi) = d$. Let $p_2 : T \to (S \otimes_R S) \otimes_S T$ be inclusion on the second factor, and let $K = \{t \in T : \phi(t) = p_2(t)\}$. We will prove the following two assertions

<u>Claim 1</u>. K is a componentially locally strongly separable extension of R.

<u>Claim 2</u>. The natural map $S \otimes_R K \to T$ is an isomorphism.

Once the claims are established we'll have that $X(S \otimes_R K)$ is isomorphic to $X(T) = A$ as $X(S \otimes_R S)$-sets, and hence that every $X(S \otimes_R S)$-set is isomorphic to an image.

The first step is to reduce to the case where R is connected. Let $x \in X(R)$. Then $K_x = \{t \in T_x : \phi_x(t) = p_{2,x}(t)\}$. If K_x is locally strongly separable over R_x and $S_x \otimes_{R_x} K_x \to T_x$ is an isomorphism for each x, then the two claims hold, so we will assume $R = R_x$ is connected. (This means that if $X = X(S_x) \subseteq X(S)$, we must replace A by $B = p^{-1}(X)$ and p by its restriction to B, and we do this. Similarly S and T are replaced by S_x and T_x.)

Let S_0, then, be a separable closure of the (connected) ring R and let $G = \mathrm{Aut}_R(S_0)$. Then $S = C(X,S_0)$ and $T = C(B,S_0)$ by III.24. Then $X(S \otimes_R S) = X\times X\times G$ (III.25) and r,d become the projections on the first and second factors, respectively. We identify $X\times X\times G X_{d,p} B$ with $X\times G\times B$ via the map $(x,y,g,b) \to (x,g,b)$ and consider the action map d under this identification.

Using the techniques of Chapter III, we get $(S \otimes_R S) \otimes_S T$ isomorphic to $C(X\times G\times B,S_0)$ by $(s_1 \otimes s_2 \otimes t)(x,g,b) = s_1(x)g(s_2(p(m))t(m))$, and, using the isomorphism and the above identification $d,\phi : C(B,S_0) \to C(X\times G\times B,S_0)$ becomes $(\phi f)(x,g,b) = f(d(x,g,m))$ while $p_2 : C(B,S_0) \to C(X\times G\times B,S_0)$ becomes $(p_2 f)(x,g,b) = g(f(b))$. So $K = \{f \in C(B,S_0) : f(d(x,g,b)) = g(f(b))\}$.

The next step is to interpret the set $X\times G\times B$. We have continuous maps $\psi : X\times G\times B \to B\times B\times G$ by $(x,g,b) \to (d(x,g,b),b,g)$ and $\delta : B\times B\times G \to X\times G\times B$ by $(b,c,g) \to (p(c),g,b)$. Then ψ followed by δ is the identity, so ψ maps $X\times G\times B$ homeomorphically onto its image H in $B\times B\times G$. Now $B\times B\times G$ (= $X(T \otimes_R T)$ by III.25) is a groupoid, and it's easy to check the properties of d (IV.29) make H a closed subgroupoid. Let $L = T^H$ (III.29); L is a locally strongly separabl extension of R by III.30, and $H = X(T \otimes_L T)$ (III.30, part 3).

We have the commutative diagram

(*)
$$\begin{array}{ccc} H & \rightrightarrows & B \\ \uparrow & & \uparrow \\ X \times G \times B & \rightrightarrows & B \end{array}$$

where: the left map = ψ , the right = 1 , the top pair are range and d projections of the groupoid H and the bottom maps are d and the pro- jection on the last factor.

Now using (*), IV.24, and the identifications we've made above, we have a commutative diagram

(**)
$$\begin{array}{ccc} T & \rightrightarrows & T \otimes_L T \\ \downarrow & & \downarrow \\ T & \rightrightarrows & (S \otimes_R S) \otimes_S T \end{array}$$

where the top pair of maps are $t \to t \otimes 1$ and $t \to 1 \otimes t$ while the bottom pair are ϕ and p_2 . The left map is the identity. T is fai- fully flat over L , and so $L = \{t \in T : t \otimes 1 = 1 \otimes t\}$. But the commutativity of (*) implies that $K = \{t \in T : \phi(t) = p_2(t)\} = L$, and hence K is locally strongly separable over R (so Claim 1 holds).

For Claim 2, we examine more closely the map δ defined above. W can identify $S \otimes_R T$ with $C(X \times G \times B, S_0)$ via $(s \otimes t)(x,g,b) = s(x)g(t($ and $T \otimes_R T$ with $C(B \times B \times G, S_0)$ by $(t_1 \otimes t_2)(b,c,g) = t_1(b)g(t_2(b))$ using the techniques of Chapters I and III. With these identifications

δ gives rise to the map $S \otimes_R T \to T \otimes_R T$ by $s \otimes t \to s \otimes t$, and, restricting to H, an isomorphism $S \otimes_R T \to T \otimes_L T$. Since the latter map is just $(S \otimes_R L \to T) \otimes_L 1_T$ and T is faithfully R-flat. $S \otimes_R L \to T$ is an isomorphism. This proves Claim 2.

To complete the proof of the theorem, we need to show that our functor is fully faithful. It is no longer necessary to assume R connected, and we return to our original ring R. The faithful part is easy: our functor $X(S \otimes_R (\cdot))$ is a composite, the first $S \otimes_R (\cdot)$ is faithful since S is faithfully R-flat and the functor $X(\cdot) : \mathcal{S}(S) \to \mathcal{P}(S)$ is faithful since it's an equivalence (IV.24). So we need to show that it's full: i.e., given T,T' componentially locally strongly separable extensions of R and an $X(S \otimes_R S)$-set map $f : X(S \otimes_R T) \to X(S \otimes_R T')$, we want an R-algebra homomorphism $h : T' \to T$ such that $X(S \otimes_R h) = f$.

Consider the maps $S \otimes_R T \rightrightarrows (S \otimes_R S) \otimes_S (S \otimes_R T)$ where the top map is $s \otimes t \to s \otimes 1 \otimes 1 \otimes t$ and the bottom $s \otimes t \to 1 \otimes s \otimes 1 \otimes t$. By IV.27 the top map gives the $X(S \otimes_R S)$-action on $X(S \otimes_R T)$. By IV.24, $f = X(k)$ for some S-homomorphism $k : S \otimes_R T' \to S \otimes_R T$. Then we have a commutative diagram

$$(***) \qquad \begin{array}{ccc} S \otimes_R T' & \rightrightarrows & (S \otimes_R S) \otimes_S (S \otimes_R T') \\ \downarrow & & \downarrow \\ S \otimes_R T & \rightrightarrows & (S \otimes_R S) \otimes_S (S \otimes_R T) \end{array}$$

where the left map is k and the right $1 \otimes_S k$. Since S is faithfully flat, the set of elements where the top maps agree is T' and where the bottom maps agree is T. Thus the restriction of k to T' gives a map $h : T' \to T$ such that $k = S \otimes_R h$ and so $f = X(S \otimes_R h)$ as desire This completes the proof that our functor is full, and hence the theorem is established.

In the context of Theorem IV.31, the topological groupoid $X(S \otimes_R S$ thus determines what the category of componentially locally strongly sep arable extensions of R looks like. (Galois theory is the theory of subobjects in this category, and we'll show in Chapter V how Galois theo works in categories of sets with groupoid action.) Since the groupoid $X(S \otimes_R S)$ plays such a fundamental role, we give it a special name:

Definition IV.32. Let R be a ring with separable closure S . Then the topological groupoid $X(S \otimes_R S)$ (with groupoid structure as in IV.28) is called the *fundamental groupoid* of R .

The set of objects of $X(S \otimes_R S)$ is $X(S)$, so if R (and hence S) is connected, $X(S \otimes_R S)$ has only one object, i.e., is a group. In this case $X(S \otimes_R S)$ is called the *fundamental group*. A set on which this group acts continuously is an $X(S \otimes_R S)$-set in the sense of IV.29, and we have the following version of IV.31 in this case:

Corollary IV.33. Let R be a connected ring with separable closure S . Then the category of locally strongly separable extensions of R i anti-equivalent to the category of profinite sets on which the fundamental group of R acts continuously.

Bibliographic Note on Chapter IV

Nearly all the material in this Chapter, in slightly different form, can be found in "The separable closure of some commutative rings", (Trans. Amer. Math. Soc. 170 (1972), 109-124) and "The fundamental groupoid of a commutative ring" (preprint). Readers consulting these sources, however, should be warned that in the first the "separable closure" discussed is more restricted than that defined here - namely, it must contain no idempotents not in the base ring. This definition was abandoned because such closures need not always exist.

Corollary IV.32 was first proved by Grothendieck in *Seminaire Geometrie Algebrique*, 1960-61, Expose V, where the idea that the essential point in Galois theory is to classify the *category* of separable algebras first appears.

V. Galois Correspondences.

We constructed, in Chapter IV, a contravariant natural equivalence between the category of componentially locally strongly separable extensions of the ring R and the category of profinite spaces surjectively over the Gleason cover of $X(R)$ on which the fundamental groupoid of R acts continuously. Problems which may be hard to solve in the first category can become tractible in the second, and in this Chapter we study one such problem: the classification of the componentially locally strongly separable subextensions of a given componentially locally strongly separable extension of R . Because of the contravariance of the equivalence, this classification of subobjects in the category of extensions becomes the classification of quotient objects in the category of spaces with groupoid action.

Let's disregard the groupoid action and topological space structures for a moment, so the problem is just classification of quotients in the category of sets. Leaving aside the categorical considerations, a quotient of a set S is (provisionally) a set T and a surjection $f : S \to T$. Now f is completely determined by the set $\mathcal{P} = \{f^{-1}(t) : t \in T\}$ of subsets of f , and $\mathcal{P}$ is a partition of S . So quotients of S correspond to partitions of S . Partitions of S correspond, in turn, to equivalence relations on S , which in turn are determined by their graphs, which are certain subsets of $S \times S$. Thus quotients of S correspond to certain subsets of $S \times S$. Now return to the category of spaces with groupoid action. The basic correspondence just outlined still obtains, as we'll see, although of course at each stage we must take care to remain in the category in question.

Finally, we relate this to the original problem of extensions, using the category equivalence: the subextensions of the extension T of R correspond to certain quotients of $T \otimes_R T$ (for if X is the space with groupoid action associated to T then subextensions of T are associated by the equivalence to quotients of X, while subobjects of $X x X$ are associated to quotients of $T \otimes_R T$).

So far, of course, all that's been said only changes the classification of subextensions of T into the classification of certain quotients of $T \otimes_R T$, and both problems appear to be equally difficult. But $T \otimes_R T$ is a componentially locally strongly separable extension, and so are the quotients of it we're interested in. If E is such a quotient, by IV.3 the kernel of $T \otimes_R T \to E$ is idempotent generated, and of course that kernel determines E. An ideal of $T \otimes_R T$ generated by idempotents determines and is determined by a certain closed subset of $X(T \otimes_R T)$. Thus we have a correspondence between subextensions of T and certain closed subsets of $X(T \otimes_R T)$.

Now we suppose further that the extension T satisfies some additional "normality" condition. Then we can show that $X(T \otimes_R T)$ is actually a topological groupoid and that the closed subsets of it corresponding to the subextensions of T are precisely the closed subgroupoids. This then gives, in its most general form, the fundamental theorem of Galois theory (and we remark its similarity to the theorem over a connected base III.30).

The main difficulty in carrying out the program just outlined is placing the arguments in a categorical (i.e., diagrammatic) setting. Readers will quickly note that while spaces with groupoid action are basically simpler objects to deal with than algebras, the unfamiliarity of the spaces requires considerable time to be spent on their trivial properties, and the reader's patience is invited.

We begin by recalling and recording the definitions of subobject, quotient object, and equivalence relation in a category. For that purpose we fix a category C. We assume that all the products, etc. in C used in the definitions exist. To avoid complications about whether monomorphisms are into or epimorphisms onto, we assume fixed classes closed under products of monomorphisms and epimorphisms of C which we refer to as special morphisms.

Definition V.1. Let B be an object of C. Two special monomorphisms $f_i : A_i \to B$, $i = 1,2$ are *equivalent* if there is an isomorphism $g : A_1 \to A_2$ such that $f_1 = f_2 g$. (It's easy to see that this is an equivalence relation.) A *subobject* of B is an equivalence class of special monomorphisms with target B. Dually, two special epimorphisms $g_i : B \to C_i$, $i = 1,2$ are *equivalent* if there is an isomorphism $h : C_1 \to C_2$ such that $g_2 = hg_1$; a *quotient object* of B is an equivalence class of special epimorphisms with source B . (We'll say: "let $f : A \to B$ be a subobject" to mean " f is a representative of a subobject of B ", and similarly for quotients.)

<u>Definition V.2.</u> Let B be an object of $\mathcal{C}$. An <u>equivalence relation in $\mathcal{C}$</u> on B is a five-tuple (E,a,b,c,d) where $a : E \to BxB$ is a subobject, $b : B \to E$, $c : E \to E$ and $d : EX_BE \to E$ are morphisms (let $pr_i : BxB \to B$ be projection on the ith factor, and let $a_i = pr_i a$ Then $EX_BE = EX_{a_1,a_2}E$.), subject to the following requirements:

1) The composite $ab : B \to BxB$ is the diagonal map.

2) If $t : BxB \to BxB$ denotes the switch map (i.e., $pr_1 t = pr_2$ and $pr_2 t = pr_1$) then $ta = ac$.

3) Let $s : (BxB)X_{pr_2,pr_1}(BxB) \to BxB$ be the canonical map (i.e., if B had elements, $s(x,y,y,z) = (x,z)$). Then $s\cdot(aX_Ba) = a$

In order to understand V.2, consider the case where $\mathcal{C}$ is the category of sets and $a : E \to BxB$ is the inclusion of the subset E . Then 1) says that $b(x) = (x,x)$ belongs to E for all x in B , 2) says that if $(x,y) \in E$ then $c((x,y)) = (y,x) \in E$ and 3) says that if $(x,y) \in E$ and $(y,z) \in E$ then $d((x,y),(y,z)) = (x,z) \in E$. In other words, E is a reflexive, symmetric, transitive relation on B - that is an equivalence relation.

If (E,a,b,c,d) is an equivalence relation on B , we call the maps $a_i = pr_i \cdot a$ the <u>associated projections</u>. As usual, we denote equivalence relations by their first entries.

Definition V.3. Let E be an equivalence relation on B with associated projections a_1, a_2. A quotient of B by E is a quotient object $f : B \to C$ of B such that: (i) $fa_1 = fa_2$ and (ii) for any morphism $g : B \to D$ such that $ga_1 = ga_2$ there is a unique morphism $h : C \to D$ such that $hf = g$.

Of course quotients don't always exist, but when they do they're unique. If $\mathcal{C}$ is the category of sets and the equivalence relation E is a subset of $B \times B$, then there is a natural projection f from B to the set of equivalence classes B/E of E on B, and f is a quotient of B by E in the sense of V.3. So if the quotient of B by E exists in our arbitrary category $\mathcal{C}$, we denote it B/E.

Now suppose the equivalence relation E on B with projections a_1, a_2 does have a quotient $f : B \to C$. Since $fa_1 = fa_2$, we have a map $p : E \to BX_C B$. We also have a map $q : BX_C B \to B \times B$, and $qp = a$.

Definition V.4. The equivalence relation E on B is effective if there is a quotient $B \to C$ of B by E such that the induced map $E \to BX_C B$ is an isomorphism.

It's an easy exercise if $\mathcal{C}$ is the category of sets, every equivalence relation is effective. There is a canonical way to build equivalence relations out of quotients:

<u>Proposition V.5.</u> Let $f : B \to C$ be a quotient object. Then BX_CB is an equivalence relation on B.

<u>Proof:</u> Let $p_i : BX_CB \to B$ be projection on the ith factor, and let $a : BX_CB \to BXB$ be such that $pr_i \cdot a = p_i$. Let $b : B \to BX_CB$ be the diagonal, $c : BX_CB \to BX_CB$ the switch map, and $d : (BX_CB)X_B(BX_CB) \to BX_CB$ the "projection on the outer factors". Then from the definition (BX_CB,a,b,c,d) is an equivalence relation on B.

<u>Definition V.6.</u> A quotient object $f : B \to C$ of B in $\mathcal{C}$ is <u>effective</u> if the quotient B/BX_CB exists and equals C.

<u>Proposition V.7.</u> The correspondences $E \to B/E$ and $C \to BX_CB$ induce inverse bijections between the class of effective quotient objects of B and the class of effective equivalence relations on B.

<u>Proof:</u> The definitions are set up in such a fashion that the proposition is a tautology.

Proposition V.7 solves, in a sense, the general classification of quotients problem. For the solution to be completely satisfactory, however, the category $\mathcal{C}$ should be such that all quotient objects and all equivalence relations are effective. We'll now show that this is the case when $\mathcal{C}$ is the category of spaces with groupoid action.

Notation V.8. Let G be a topological groupoid whose underlying set is profinite. Then $M(G)$ denotes the category of profinite spaces, surjectively over the space of objects of G, on which G acts continuously (see Definition IV.29). The special monomorphisms of $M(G)$ are the injections and the special epimorphisms are the surjections.

Lemma V.9. Let C be the category of profinite spaces. Then every equivalence relation in C has a quotient, and all quotient objects and equivalence relations in C are effective.

Proof: As remarked above, the Lemma is true if C is the category of sets. To show it for the category of profinite spaces, we begin by checking that the set-theoretic quotient of a profinite equivalence relation E on a profinite space B carries a natural profinite space structure. We can identify BxB with $BxBx\{1\}$ (here $\{1\}$ is the one-element group) and hence BxB can be regarded as a profinite groupoid (III.29), the identities being the pairs (b,b), the inverse of (b,b') being (b',b), and the product $(b,b')(b',b'')$ being (b,b''). Thus E, regarded as a subspace of BxB, is a subgroupoid. Now write $B = \text{proj lim } B_i$ where each B_i is finite, $p_i : B \to B_i$, is a continuous surjection, and the partition $\{p_i^{-1}(x) : x \in B_i\}$ is E-homogeneous (in the sense of III) for each i (this can be done by III.31). It follows that, for each i, $E_i = (p_i x p_i)(E) \subseteq B_i x B_i$ is an equivalence relation on B_i (see III.31), and we have compatible surjections of the (topological) quotients $q_i : B/E \to B_i/E_i$ and hence a continuous surjection $q : B/E \to \text{proj lim } (B_i/E_i)$. Since E is closed in $B \times B$, we have $E = \bigcap (p_i \times p_i)^{-1}(E_i)$ and hence q is injective, and thus a homeomorphism. So B/E is profinite, and B has a quotient by E in $\mathcal{C}$.

To complete the proof of the Lemma, we note that if $B \to C$ is a quotient object in profinite spaces, then $B \times_C B$, as profinite spaces, has the same underlying set as the set-theoretic fibre product. We just saw that quotients have the same underlying set as set-theoretic quotients. Equivalence relations and quotients are effective in $\mathcal{C}$ if certain continuous maps are bijections. Since, as already noted, the maps are bijective set-theoretically, they're bijections in $\mathcal{C}$ and the Lemma obtains.

Of course one need not appeal to groupoids to prove V.9; we do so here only to emphasize the connection between the groupoids and equivalence relations (see also V.19 below).

Proposition V.10. Let $M(G)$ denote the category of V.8. Then every equivalence relation in $M(G)$ has a quotient, and all quotient objects and equivalence relations in $M(G)$ are effective.

Proof: Let Z be the space of objects of G , let (B,p) be an object of $M(G)$ and let (E,q) in $M(G)$ be an equivalence relation on (B,p) Then E is a subspace of $(B,p) \times (B,p) = B \times_Z B$, on which G acts, which satisfies the conditions of V.2. This means that E , considered as a (closed) subspace of BxB , is an equivalence relation on B in the category of profinite spaces. We'll show that G acts on the (profinite space) quotient B/E in such a way that B/E is the quotient of (B,p) by (E,q) . First, if a_1 and a_2 are the associated projections of (E,q) , since they're maps in $M(G)$, we must have $pa_i = q$, $i = 1,2$. Thus $p : B \to Z$ induces a continuous surjection $k : B/E \to Z$, and the quotient map $(B,p) \to (B/E,k)$ is a map of spaces over Z . Let $b \in B$ have image $[b]$ in B/E , and suppose $g \in G$ is such that $\text{domain}(g) = p(b) = k([b])$. Let $a(g,[b])$ be the image of gb in B/E . If $(b',b) \in E$, then we claim that gb and gb' have the same image in B/E . For $\text{domain}(g) = q(b,b') = pa_1(b) = pa_2(b')$ and hence $g(b,b') = (gb,gb')$ belongs to E . Thus the relation a is a well-defined function and the diagram

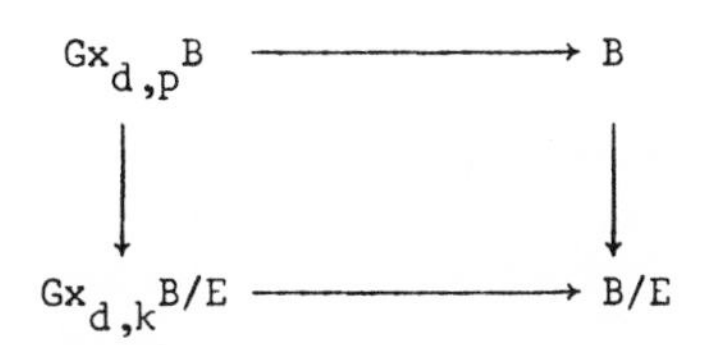

where the top map is the action of G on B , the bottom is a and the side maps come from the quotient map, commutes. It follows that a gives an action of G on B/E in the sense of IV.29. Finally $(B/E,k)$ is the quotient of (B,p) by (E,q) in $M(G)$ since B/E is the profinite space quotient of B by E .

Now suppose $(B,p) \to (C,\ell)$ is a quotient object in $M(G)$. Then $(B,p)X_{(C,\ell)}(B,p) = (BX_C B, p\cdot pr_1)$ and so the underlying space of the equivalence relation corresponding to the quotient object is the relation corresponding to the quotient object $B \to C$ in profinite spaces. Thus exactly as in the proof of V.9, equivalence relations and quotients in $M(G)$ are effective since the corresponding result is true for the category of profinite spaces.

Corollary V.11. Let $M(G)$ denote the category of V.8, and let (B,p) be an object of $M(G)$. Then the correspondences $(E,q) \to (B,p)/(E,q)$ and $(C,\ell) \to (B,p)X_{(C,\ell)}(B,p)$ induce inverse bijections between the set of equivalence relations on (B,p) and the quotients (C,ℓ) of (B,p).

Proof: We interpret V.7 in the light of V.10.

The above corollary applies in particular when G is the fundamental groupoid of a ring R. In this case the category $M(G)$ is dual, by IV.31, to the category of componentially locally strongly separable extensions of R. We want to dualize the notion of equivalence relation so as to be able to phrase V.11 entirely in terms of this latter category We'll call the dual of an equivalence relation an equivalence correlation but only describe the notion in detail for categories of algebras.

Definition V.12. Let R be a ring and C a category of R-algebras. Let S be an R-algebra. An equivalence correlation on S in C is a five-tuple (T,a,b,c,d) where $a : S \otimes_R S \to T$ is an onto homomorphism, $b : T \to S$, $c : T \to T$ and $d : T \to T \otimes_S T$ are homomorphisms (let

$q_i : S \to S \otimes_R S$ be inclusion on the ith factor and let $a_i = aq_i$.
Then $T \otimes_S T = T \otimes_{a_1, a_2} T$), and such that the following all commute:

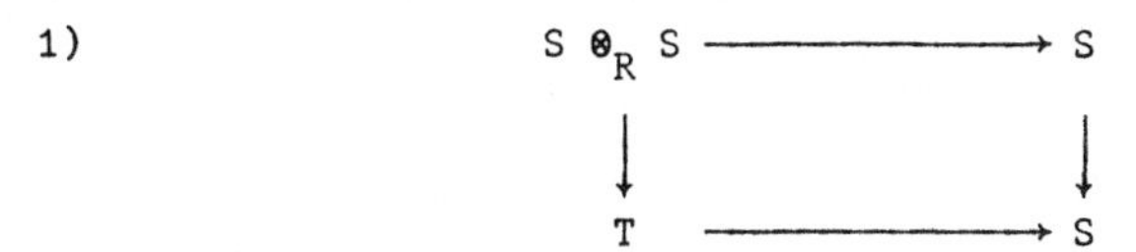

where the top map is a multiplication, the bottom a , the left b , and the right 1 ;

2)
$$\begin{array}{ccc} S \otimes_R S & \longrightarrow & S \otimes_R S \\ \downarrow & & \downarrow \\ T & \longrightarrow & T \end{array}$$

where the vertical maps are a , the bottom c , and the top $x \otimes y \to y \otimes x$;

3)
$$\begin{array}{ccc} S \otimes_R S & \longrightarrow & (S \otimes_R S) \otimes_S (S \otimes_R S) \\ \downarrow & & \downarrow \\ T & \longrightarrow & T \otimes_S T \end{array}$$

where the left map is a , the right $a \otimes a$, the bottom d and the top $x \otimes y \to x \otimes 1 \otimes 1 \otimes y$.

The correlation is componentially locally strongly separable if T is over R .

Using the language of V.12 and the remarks following V.11, we have:

Proposition V.13. Let R be a ring and S a componentially locally strongly separable extension of R. Then there is a one-one correspondence between componentially locally strongly separable subextensions of R in S and componentially locally strongly separable equivalence correlations on S, given as follows:

i) If $R \subset T \subset S$ and T is componentially locally strongly separable, then the associated equivalence relation is $(S \otimes_T S, a, b, c, d)$ where $a : S \otimes_R S \to S \otimes_T S$ is the canonical map, $b : S \otimes_T S \to S$ is multiplication, $c : S \otimes_T S \to S \otimes_T S$ is the switch map and $d : S \otimes_T S \to (S \otimes_T S) \otimes_S (S \otimes_T S)$ sends $x \otimes y$ to $x \otimes 1 \otimes 1 \otimes y$.

ii) If (B,a,b,c,d) is a componentially locally strongly separable equivalence correlation on S, the corresponding subalgebra is $\{x \in S : a(1 \otimes x) = a(x \otimes 1)\}$.

Proof: The dual of this was shown in V.11, and by IV.31 this makes the Proposition itself hold.

Without additional information about the extension S, it is hard to improve on the correspondence of V.13. Just as for the case where the ring R is connected (see Chapter III), what's needed is a suitable definition of normality.

Definition V.14. Let S be a componentially locally strongly separable extension of the ring R. S is said to be C-normal if the natural

maps $X((S \otimes_R S) \otimes_S (S \otimes_R S)) \to X(S \otimes_R S) \times_{X(S)} X(S \otimes_R S)$ and
$X((S \otimes_R S) \otimes_S (S \otimes_R S) \otimes_S (S \otimes_R S)) \to X(S \otimes_R S) \times_{X(S)} X(S \otimes_R S) \times_{X(S)} X(S \otimes_R S)$
are homeomorphisms.

We'll see shortly the usefulness of V.14. First we compare it with the normality defined in III.17:

Lemma V.15. Let R be a connected ring and S a locally strongly separable extension of R. Then if S is locally weakly Galois (III.21), S is C-normal.

Proof: By III.24, $S = C(Y,S_0)$ where Y is a profinite space and S_0 an infinite Galois (III.2) R-algebra. Let $G = Aut_R(S_0)$. Then by III.25, $X(S \otimes_R S) = Y \times Y \times G$. Thus $X(S \otimes_R S) \times_{X(S)} X(S \otimes_R S)$ can be identified with $\{((a,b,g),(b,c,h)) : a,b,c \in Y,\ g,h \in G\}$, and this latter can be identified with $Y \times Y \times Y \times G \times G$. Now also $(S \otimes_R S) \otimes_S (S \otimes_R S)$ is isomorphic to $S \otimes_R S \otimes_R S$ via $x \otimes y \otimes z \otimes w \to x \otimes yz \otimes w$, and
$S \otimes_R S \otimes_R S = C(Y \times Y \times G, S_0) \otimes_R C(Y,S_0) = C(Y \times Y \times G \times Y, S_0 \otimes_R S_0) = C(Y \times Y \times Y \times G \times G, S_0)$
(for the last, we used $S_0 \otimes_R S_0 = C(G,S_0)$). Thus $X(S \otimes_R S \otimes_R S) = Y \times Y \times Y \times G \times G$. We leave it to the reader to check that these identifications are all compatible, i.e., we have $X((S \otimes_R S) \otimes_S (S \otimes_R S)) \to X(S \otimes_R S \otimes_R S) \to Y^{(3)} \times G \to X(S \otimes_R S) \times_{X(S)} X(S \otimes_R S)$ is the natural map. The proof that the other natural map is a homeomorphism is similar.

Proposition V.16. Let R be any ring and S an R-algebra such that, for each x in $X(R)$, S_x is a locally weakly Galois extension of R. Then S is C-normal.

Proof: Of course S is componentially locally strongly separable over R. We have, for each x in $X(R)$, a commutative diagram

$$\begin{array}{ccc} X((S_x \otimes_{R_x} S_x) \otimes_{S_x} (S_x \otimes_{R_x} S_x)) & \longrightarrow & X((S \otimes_R S) \otimes_S (S \otimes_R S)) \\ \downarrow & & \downarrow \\ X(S_x \otimes_R S_x) X_{X(S_x)} X(S_x \otimes_{R_x} S_x) & \longrightarrow & X(S \otimes_R S) X_{X(S)} X(S \otimes_R S) \end{array}$$

and by V.15 the left map is a homeomorphism. Call the top and bottom maps f_x and g_x, respectively, and note that both are injections. Further, any element of the upper right vertex is in the image of some f_x, and similarly for the lower right vertex and the g_x's. Call the right hand map h. It follows that h is surjective. If $h(z) = h(w)$ z and w are in the image of the same f_x, and it follows that $z = w$ So h is a homeomorphism. The proof that the other natural map is also a homeomorphism is similar.

Proposition V.17. Let R be any ring and S a componentially locally strongly separable C-normal extension of R. Then $X(S \otimes_R S)$ is, canonically, a topological groupoid.

Proof: Let $m : S \otimes_R S \to S$ be multiplication, $i : S \otimes_R S \to S \otimes_R S$ the switch map and $d : S \otimes_R S \to (S \otimes_R S) \otimes_S (S \otimes_R S)$ be given by $d(x \otimes y) = x \otimes 1 \otimes 1 \otimes y$. Consider the five-tuple

$(X(S \otimes_R S), X(S), X(m), X(i), d^*)$, where $d^* = X(d)$ preceded by the homeomorphism $X((S \otimes_R S) \otimes_S (S \otimes_R S)) \to X(S \otimes_R S)$ of V.14. We claim that this five-tuple is a groupoid, and verify first that d^* is associative. The diagram

$$\begin{array}{ccc} S \otimes_R S & \longrightarrow & (S \otimes_R S) \otimes_S (S \otimes_R S) \\ \downarrow & & \downarrow \\ (S \otimes_R S) \otimes_S (S \otimes_R S) & \longrightarrow & (S \otimes_R S) \otimes_S (S \otimes_R S) \otimes_S (S \otimes_R S) \end{array}$$

where the top map is d , the bottom map $1 \otimes_S d$, the left map d , and the right map $d \otimes_S 1$, commutes. Thus, applying $X(\cdot)$, $X(d \otimes_S 1)X(d) = X(1 \otimes_S d)X(d)$. Since S is C-normal, this last equation implies $(d^* X_{X(S)} 1)d^* = (1 X_{X(S)} d^*)d^*$, so d^* is associative. The verification of the other parts of III.26 are similar and left to the reader.

We'll need the following lemma in the proof of Theorem V.19 below.

<u>Lemma V.18</u>. Let R be a ring and S a C-normal componentially locally strongly separable extension of R . Let E be a R-algebra and $p : S \otimes_R S \to E$ a surjection. Let $p_1(s) = p(s \otimes 1)$ and $p_2 = p(1 \otimes s)$ map S to E , and let $E \otimes_S E = E \otimes_{p_2, p_1} E$. Then the natural map $X(E \otimes_S E) \to X(E) X_{X(S)} X(E)$ is a bijection.

<u>Proof</u>: We have a commutative diagram

$$\begin{array}{ccc} X(E \otimes_S E) & \longrightarrow & X((S \otimes_R S) \otimes_S (S \otimes_R S)) \\ \downarrow & & \downarrow \\ X(E) X_{X(S)} X(E) & \longrightarrow & X(S \otimes_R S) X_{X(S)} X(S \otimes_R S) \end{array}$$

where the right map is a homeomorphism, and both horizontal maps are injections. Thus the left map, h , is an injection. Suppose z,w are in $X(E)$ with $X(p_1)(z) = X(p_2)(w) = a$. Choose x in $X(E_z \otimes_{S_a} E_w)$, and define y in $X(E \otimes_S E)$ by $(E \otimes_S S)_y = (E_z \otimes_{S_a} E_w)_x$. Then $h(y) = (z,w)$, so h is also surjective.

Now we use V.13 and V.17 to prove the fundamental theorem of Galois theory in its general form.

Theorem V.19. Let R be a ring and S a C-normal componentially locally strongly separable extension of R . Then there is a one-one correspondence between componentially locally strongly separable subextensions T of S and closed subgroupoids H of $X(S \otimes_R S)$ (see V.17 for the groupoid structure) constructed as follows:

i) To the extension T corresponds the subgroupoid which is the image of $X(S \otimes_T S) \to X(S \otimes_R S)$.

ii) To the subgroupoid H corresponds the subalgebra $\{s \in S : (s \otimes 1)_z = (1 \otimes s)_z$ for all $z \in H\}$.

Proof: We already know (V.13) that componentially locally strongly separable subextensions correspond to equivalence correlations. To prove the theorem, we'll show that subgroupoids and equivalence correlations also correspond.

Thus let $X(S \otimes_R S)$ carry the groupoid structure of V.17 and let H be a closed subgroupoid of it. Choose a family e_i of idempotents of $S \otimes_R S$ such that the open set $X(S \otimes_R S) - H$ is the union of the $N(e_i)$. Let I be the ideal of $S \otimes_R S$ generated by the e_i, and let $B = S \otimes_R S/I$. Then the canonical injection $X(B) \to X(S \otimes_R S)$ has image H. We're going to show that B is (the underlying quotient object of) an equivalence correlation on S.

First, H contains all the identities. The identities of $X(S \otimes_R S)$ are the image of the map $X(S) \to X(S \otimes_R S)$ induced from multiplication. Thus, writing $\overline{a \otimes b}$ for the image of $a \otimes b \to S \otimes_R S$ in B, the map $m : B \to S$ by $\overline{a \otimes b} = ab$ is well-defined. Next, H is closed under inverison. This means that for $a,b \in S$ and $z \in H$, $(a \otimes b)_z = (b \otimes a)_z$. Thus the map $t : B \to B$ given by $t(\overline{a \otimes b}) = \overline{b \otimes a}$ is well-defined. Finally, H is closed under the partial multiplication. Using V.18 to identify $HX_{X(S)}H$ and $X(B \otimes_S B)$, this means that the map $g : B \to B \otimes_S B$ given by $g(\overline{a \otimes b}) = (\overline{a \otimes 1}) \otimes_S (\overline{1 \otimes b})$ is well-defined. Then the five-tuple (B,p,m,t,g), where $p : S \otimes_R S \to B$ is the canonical map and the rest are as just defined, is an equivalence correlation on B (the maps are set up so the diagrams 1), 2), 3) of V.12 all commute) which is componentially locally strongly separable over R since $S \otimes_R S$ is and I is generated by idempotents (see IV.4).

Now let (B,a,b,c,d) be a componentially locally strongly separable equivalence correlation on S. Use V.18 to identify $X(B)X_{X(S)}X(B)$ with $X(B \otimes_S B)$, and let d^* be $X(d)$ preceded by this identification. Then $(X(B),X(b),X(c),d^*)$ is a groupoid and

$X(a) : X(B) \to X(S \otimes_R S)$ is an injection of $X(B)$ onto a closed subgroupoid of $X(S \otimes_R S)$. (To see that $X(B)$ is a groupoid, we apply $X(\cdot)$ to the diagrams of V.12 to see that the maps $X(b), X(c), d^*$ are compatible with the groupoid structure on $X(S \otimes_R S)$ defined in V.17.) Finally, since B is componentially locally strongly separable over R the kernel of a , call it I , is idempotent generated (IV.3).

Thus subgroupoids and equivalence correlations correspond. To see that this correspondence is one-one, note that the ideal I generated by idempotents such that $X(S \otimes_R S/I)$ is the subgroupoid H is uniquely determined by H , and the ideal I generated by idempotents which is the kernel of the equivalence correlation $S \otimes_R S \to B$ is uniquely determined by the correlation.

Now to complete the proof of the theorem, to the componentially locally strongly separable subextension $T \subsetneq S$ corresponds the equivalence correlation $S \otimes_T S$ to which corresponds the subgroupoid $X(S \otimes_T S)$. Then to the subgroupoid H corresponds the equivalence relation $B = (S \otimes_R S)/I$ and to it corresponds the subalgebra $\{s \in S : s \otimes 1 \equiv 1 \otimes s \pmod{I}\}$. It's easy to check that under our conditions, $s \otimes 1 \equiv 1 \otimes s \pmod{I}$ if and only if $(s \otimes 1)_x = (1 \otimes s)_x$ for all x in H .

Theorem V.19 applies, of course, to any ring. It's an instructive exercise, however, to see that, in case R is connected, it reduces to the Galois theory of III.30.

Bibliographic note for Chapter V

The notion of equivalence relation in a category is Grothendieck's, and appears at various points in the Seminaire Geometrie Algebrique - wherever the techniques of descent or passage to the quotient occur. One such place is Gabriel's "Construction de préschemas quotients" in the SGA 1963-1964, fasc. 2a, exposé 5.

The rest of Chapter V appears here for the first time. A Galois theorem for componentially locally strongly separable extensions was announced, without proof, in "The separable closure of some commutative rings"; the proof envisioned was a straightforward generalization of the Galois theorem for locally strongly separable algebras proved in "Galois groupoids".

Bibliography

Note: Each chapter ends with a bibliographic note explaining the sources of the material in that chapter.

DeMeyer, F. and E. Ingraham, Separable Algebras over Commutative Rings, Math. Lecture Notes #181, Springer-Verlag, New York, 1971.

Gabriel, P., Construction de préschémas quotient, Schémas en Groupes (Sém. Géométrie Algébrique, Inst. Hautes Études Sci., 1963/64) fasc. 2a, exposé 5.

Gleason, A., "Projective topological spaces," Ill. J. Math. 2(1959), 482-489.

Higgins, P., Categories and Groupoids, Van Nostrand Reinhold Math. Studies 32, London, 1971.

Hurewicz, W. and H. Wallman, Dimension Theory, Princeton University Press, Princeton, 1948.

Magid, A., The Fundamental Groupoid of a Commutative Ring, preprint.

________, "Galois groupoids," J. of Algebra 18(1971), 89-102.

________, "Pierce's representation and separable algebras," Ill. J. of Math. 15(1971), 114-121.

________, "The separable closure of some commutative rings," Trans. Amer. Math. Soc. 170(1972), 109-124.

Pierce, R., "Modules over commutative regular rings," Mem. Amer. Math. Soc. No. 70(1967).

Villamayor, O. and D. Zelinsky, "Galois theory for rings with finitely many idempotents," Nagoya Math. J. 27(1966), 721-731.

________________________, "Galois theory with infinitely many idempotents," Nagoya Math. J. 35(1969), 83-98.

Wilansky, A., Topology for Analysis, Ginn, Waltham, Massachusetts, 1970.

INDEX

A

$Alg_R(A,B)$, 55
  standard topology of, 58
$Aut_R(S)$, 28

B

B(R), 28
Boolean ideal, 27
Boolean spectrum, 24

C

C(X,Y), 19
C-normal extension, 122
Comp (X), 6
Componentially strongly separable, 79
Componentially locally strongly separable, 79
  characterization of, 81
  transitivity of, 84

E

e(S/R), 53
Equivalence relation, 114
  effective, 115
  quotient by, 115
Equivalence correlation, 120
Extension of rings, 45

F

Fundamental group, 108
Fundamental groupoid, 108
Fundamental theorem (connected), 69
Fundamental theorem (general), 126

G

Galois extension, 46
  infinite Galois extension, 46
Gleason cover, 13
Groupoid, 67
  finite, 67
  profinite, 67
  subgroupoid, 67
Groupoid-set, 102

L

Locally strongly separable extension, 45
Locally weakly Galois extension, 63

M

Minimal extension, 92
Minimal map, 9

N

$N_R(e)$, N(e), 29
Normal extension, 60
  C-normal, 122

P

Partition, 3
Profinite group action, 15
  effective, 16
Profinite space, 1
Projective cover, 13

Q

Quasi-component, 7

Quasi-covering projection, 8
Quotient object, 113
Quotient of equivalence relation, 115

S

Separable closure (connected), 46
existence, 51
Separable closure (general), 93
existence, 93
Separable extension, 45
componentially locally strongly separable, 79
locally strongly separable, 45
separable componentially strong, 82
strongly separable, 45
Separability idempotent, 53
Separably closed (connected), 46
Separably closed (general), 90
Separated subspaces, 1
Space of connected components, 6
Space of functions, 19
Strongly separable extension, 45
Subgroupoid, 68
Subobject, 113

T

Totally disconnected space, 4

W

Weakly Galois extension, 63
Weakly uniform ring, 42

X

X(R), 24